The Gatlings
at Santiago

The Gatlings at Santiago

The History of the Gatling Gun
Detachment, U. S. Fifth Army Corps,
During the Spanish-American War,
Cuba, 1898

John H. Parker

LEONAUR

The Gatlings at Santiago: the History of the Gatling Gun Detachment,
U. S. Fifth Army Corps, During the Spanish-American War, Cuba, 1898
by John H. Parker

Leonaur is an imprint of Oakpast Ltd

ISBN 978-1-84677-912-1 (hardcover)
ISBN 978-1-84677-911-4 (softcover)

http://www.leonaur.com

Publisher's Notes

The opinions of the authors represent a view of events in which he
was a participant related from his own perspective,
as such the text is relevant as an historical document.

The views expressed in this book are not necessarily
those of the publisher.

Contents

To the Enlisted Members
of the Detachment. . . .

Who, by Their Devotion,
Courage and Endurance,
Made Its Success Possible,
this Volume is Dedicated as a
Token of Esteem by the Author

Authors's Note

The photographic illustrations in this work are due to the courage and kindness of Mr. John N. Weigle, of Gettysburg, Pa. This young man was first sergeant of the Gatling Gun Detachment, and took with him a large supply of material. It was his delight to photograph everything that occurred, and his pleasure to furnish a set of photographs for the use of the author. Mr. Weigle was recommended for a commission in the Regular Army of the United States, for his extreme gallantry in action, and is a magnificent type of the American youth. The thanks of the author are tendered to him for the photographic illustrations so generously supplied.

Preface

On the morning of July 1st, the dismounted cavalry, including my regiment, stormed Kettle Hill, driving the Spaniards from their trenches. After taking the crest, I made the men under me turn and begin volley-firing at the San Juan Blockhouse and entrenchments against which Hawkins' and Kent's Infantry were advancing. While thus firing, there suddenly smote on our ears a peculiar drumming sound. One or two of the men cried out:

"The Spanish machine guns!"

But, after listening a moment, I leaped to my feet and called, "It's the Gatlings, men! It's our Gatlings!"

Immediately the troopers began to cheer lustily, for the sound was most inspiring. Whenever the drumming stopped, it was only to open again a little nearer the front. Our artillery, using black powder, had not been able to stand within range of the Spanish rifles, but it was perfectly evident that the Gatlings were troubled by no such consideration, for they were advancing all the while.

Soon the infantry took San Juan Hill, and, after one false start, we in turn rushed the next line of block-houses and entrenchments, and then swung to the left and took the chain of hills immediately fronting Santiago. Here I found myself on the extreme front, in command of the fragments of all six regiments of the cavalry division. I received orders to halt where I was, but to hold the hill at all hazards.

The Spaniards were heavily reinforced and they opened a tremendous fire upon us from their batteries and trenches. We laid down just behind the gentle crest of the hill, firing as we got the chance, but, for the most part, taking the fire without responding.

As the afternoon wore on, however, the Spaniards became bolder, and made an attack upon the position. They did not push it home, but they did advance, their firing being redoubled.

We at once ran forward to the crest and opened on them, and, as we did so, the unmistakable drumming of the Gatlings opened abreast of us, to our right, and the men cheered again. As soon as the attack was definitely repulsed, I strolled over to find out about the Gatlings, and there I found Lieut. Parker with two of his guns right on our left, abreast of our men, who at that time were closer to the Spaniards than any others.

From thence on, Parker's Gatlings were our inseparable companions throughout the siege. They were right up at the front. When we dug our trenches, he took off the wheels of his guns and put them in the trenches. His men and ours slept in the same bomb-proofs and shared with one another whenever either side got a supply of beans or coffee and sugar. At no hour of the day or night was Parker anywhere but where we wished him to be, in the event of an attack. If a troop of my regiment was sent off to guard some road or some break in the lines, we were almost certain to get Parker to send a Gatling along, and, whether the change was made by day or by night, the Gatling went. Sometimes we took the initiative and started to quell the fire of the Spanish trenches; sometimes they opened upon us; but, at whatever hour of the twenty-four the fighting began, the drumming of the Gatlings was soon heard through the cracking of our own carbines.

I have had too little experience to make my judgment final; but certainly, if I were to command either a regiment or a brigade, whether of cavalry or infantry, I would try to get a Gatling battery—under a good man—with me. I feel sure that the greatest possible assistance would be rendered, under almost all circumstances, by such a Gatling battery, if well handled; for I believe that it could be pushed fairly to the front of the firing-line. At any rate, this is the way that Lieut. Parker used his battery when he went into action at San Juan, and when he kept it in the trenches beside the Rough Riders before Santiago.

Theodore Roosevelt

CHAPTER 1

L'Envoi

The history of the Gatling Gun Detachment, Fifth Army Corps, is to a certain extent the history of the Santiago campaign. The detachment was organized on the spur of the moment, to utilize material which would otherwise have been useless, and was with the Fifth Corps in all the campaign. It participated in all the fighting of that campaign, except the fight at La Guasimas, and was disbanded upon the return of the Fifth Corps to Montauk. Whatever hardships were endured by the Fifth Corps were shared by this detachment; whatever dangers were faced by the Fifth Corps were faced by it also; where the hottest fighting occurred this detachment went in and stayed; and at the surrender it was paraded, to use the words of General Shafter, "Upon that portion of the line which it occupied so promptly and defended so well."

But this memoir is not intended as a history of that campaign nor of the Fifth Corps. The author has not the data available to cover so large a field, nor the ability to do justice to the courage, fortitude, and endurance so heroically displayed by that gallant army. That story will be written by abler pens, and will be the wonder of the world when it is told. This story is that of an experiment. It is told to lay before the general public, as well as the military critic, the work of a little detachment of thirty-seven men, armed with an untried weapon, organized in the short space of four days preceding July 1, 1898, and which without proper equipment, adequate instruction, or previous training, in the face of discouragements and sneers, and in spite of obstacles enough to make the mere retrospect sickening, still achieved

CUBANS (?)

GEN. CHAFFEE

GEN. McKIBBENS

GEN. LUDLOW

Cobre Road

Wire Entanglements

Cuabitas Road

GEN. LAWTON

Battery

Harbor of Santiago

CITY OF
SANTIAGO

6 in. M. Gun

Battery

Santa Cruz

GEN. WOOD

Road

Tit FORT

B.

GEN. SUMNER

Wire

Entanglements

GEN. PEARSON

Abandoned American Trenches

S. Juan B. H.

Ligh

Path

GEN. BATES

San Juan River

Moro Mesa.

Ligl

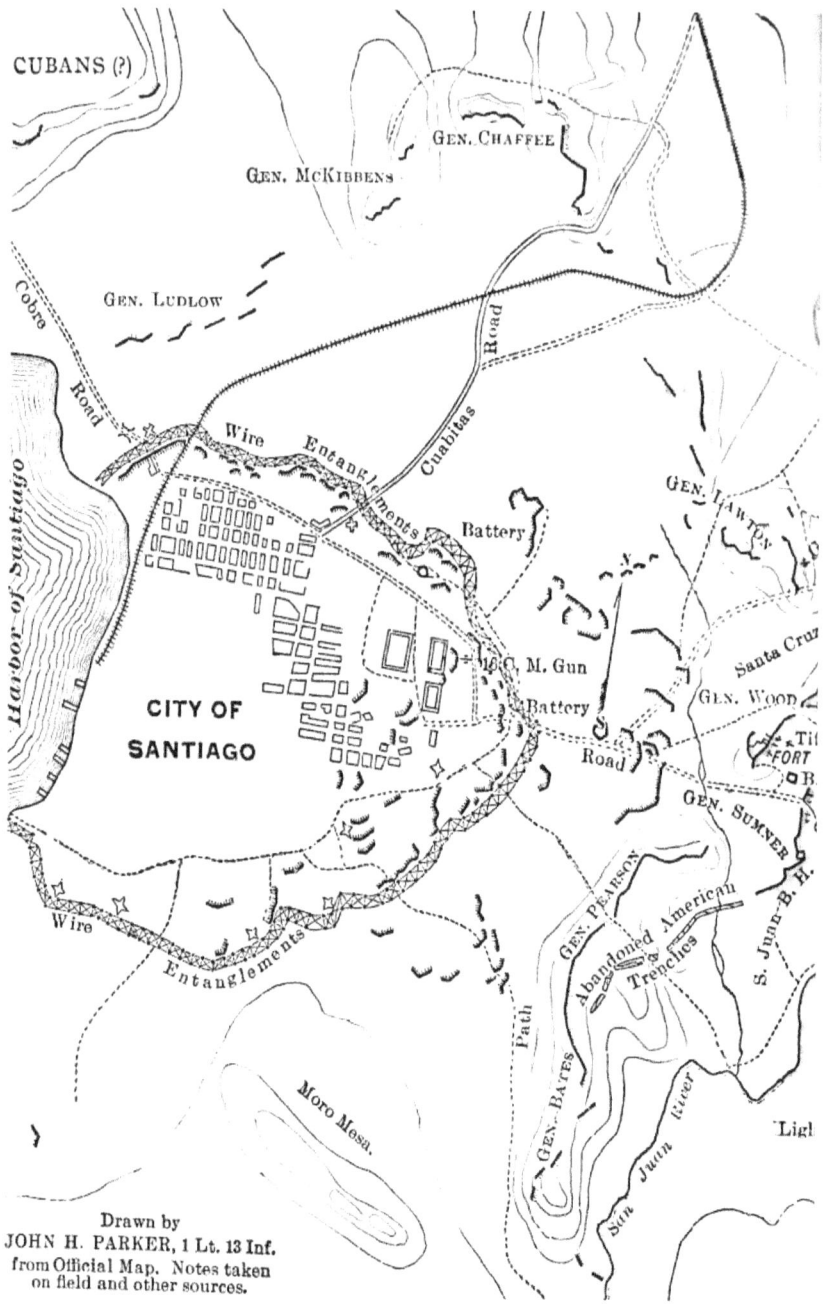

Drawn by
JOHN H. PARKER, 1 Lt. 13 Inf.
from Official Map. Notes taken
on field and other sources.

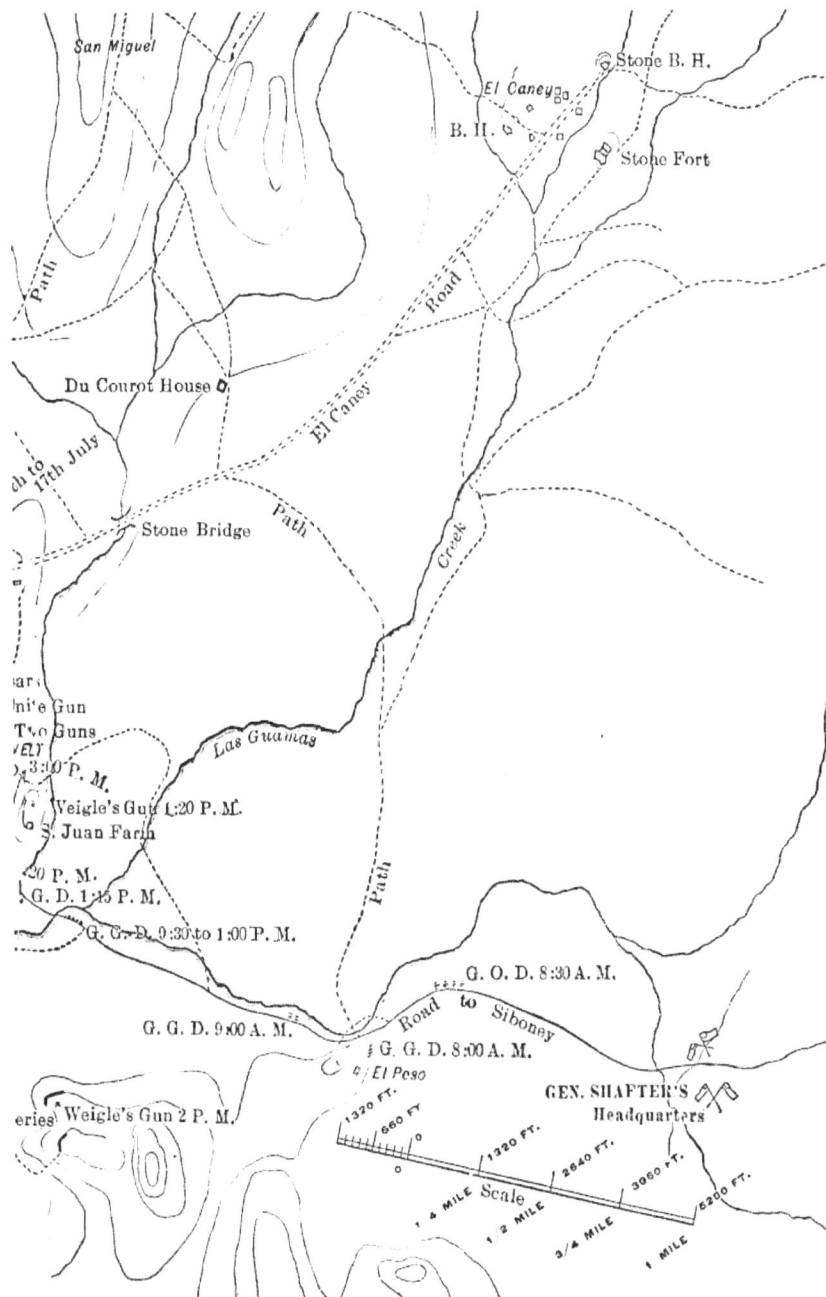

San Miguel

Stone B. H.

El Caney

B. H.

Stone Fort

Road

El Caney

Du Courot House

Path

17th July

Stone Bridge

Path

Creek

Las Guamas

Ini'e Gun

Two Guns

VELY

3:00 P. M.

Veigle's Gun 1:20 P. M.

S. Juan Farm

20 P. M.

G. D. 1:15 P. M.

G. C. D. 9:30 to 1:00 P. M.

Path

G. O. D. 8:30 A. M.

G. G. D. 9:00 A. M.

Road to Siboney

G. G. D. 8:00 A. M.

El Peso

eries Weigle's Gun 2 P. M.

GEN. SHAFTER'S
Headquarters

1320 FT.

660 FT.

1320 FT.

2640 FT.

3060 FT.

6200 FT.

1 4 MILE Scale

1 2 MILE

3 4 MILE

1 MILE

for itself a warm place in the hearts of all true soldiers, and covered itself with glory upon the hardest fought battle-field of the Hispano-American War.

This story is to commemorate the gallantry of the enlisted men who helped to make history and revolutionize tactics at Santiago. It will tell of the heroism of the plain American Regular, who, without hope of preferment or possibility of reward, boldly undertook to confute the erroneous theories of military compilers, who, without originality or reason, have unblushingly cribbed the laboured efforts of foreign officers, and foisted these compilations of second-hand opinions upon the American Army as military text-books of authority and weight. These literary soldiers declared, following the lead of their foreign guides, that "The value of machine guns on the battle-field is doubtful," and that "Their offensive value is probably very small." They also agreed, with most touching unanimity, that "A direct assault upon a fortified position, occupied by good, unshaken infantry, armed with the modern rifle and plentifully supplied with ammunition is sure to fail, unless made by overwhelming numbers and prepared by strong and accurate fire by artillery."

These servile imitators of foreign pen soldiers were destined to see all their pet theories exploded by the grim old mountain puma from California and his brave Fifth Corps. They were to learn, so far as they are capable of learning, that the American Regular makes tactics as he needs them; that the rules of war established by pen soldiers do not form the basis of actual operations in the field; that theories must go to the wall before the stern logic of irrefutable facts; and that deductions based on the drill-made automatons of European armies are not applicable to an army composed of American Volunteer Regulars, led by our trained officers.

We shall see that an army destitute of cavalry, and hence without "eyes"; not supported by artillery; in the most difficult country over which soldiers ever operated, and without maps or reconnaissance—in twenty days shut up and captured an army of twice its own effective strength, in a strongly fortified city, with better served and more numerous artillery.

We shall find that when the "sledge" was not at hand, American ingenuity was able to use the "mallet" instead, making light machine guns perform all the function of artillery, and dispensing altogether, so far as any practical results were concerned, with that expensive and much overrated arm; that the Regular private is capable of meeting all demands upon his intelligence, and that the American non. com. is the superior of foreign officers.

It is also hoped to place before the intelligent American public some correct ideas of the new arm which was tried thoroughly at Santiago for the first time in the history of the world. The machine gun is the latest practical product of American inventive genius applied to war. The first form of this weapon tried, the *mitrailleuse*, was not very successful. It failed, not on account of faults of construction, or imperfect mechanism, but because its proper tactical employment had not been thought out by the French army. Since that time machine guns have been greatly improved, but no one has succeeded in making their great value appreciated by military authorities. The failures of the French brought the gun into disfavour, and created a prejudice against its employment.

The Artillery of the world, which poses in every country as an elite body of scientific fighters, and is often found on the battlefield to be an aggregation of abstruse theorists, were jealous and contemptuous. They said, "See how easily the artillery knocked out machine guns at Gravelotte." The Cavalry of the world, famous everywhere for an *esprit-de-corps* which looks haughtily down on all other arms of the service, were too deeply absorbed in the merits of sabre *vs.* revolver, and in the proper length of their spectacular plumes, to give a second thought to this new, untried, and therefore worthless weapon. The world's Infantry, resting upon the assumption that it is the backbone of all armies, and the only real, reliable fighting body under all conditions, left the consideration of these vague dreams of mechanical destructiveness to lunatics, cranks, and philanthropists.

In our own country the Ordnance Department, which is the trial court before which all military inventions must appear, scouted the idea of usefulness of machine guns even after war was declared, and adhered to the view that machine guns, in the very nature of things, could never be useful except in the defence of fortified positions; that they never could be brought up on the battlefield, nor used if they were brought up. This view was that of a prominent young officer of that department who wrote a report on the subject, and it seemed to express the views of the department.

This view must have been that of our War Department, for it did not even acknowledge the receipt of drawings and specifications for a machine gun carriage, offered freely to the Government as a gift by the inventor six months before the war, together with the first correct tactical outline of the proper use of machine guns ever filed in any War Office in the world. This invention was designed to facilitate the use of the machine gun by making its advance with the skirmish line possible on the offensive, and was recommended by the whole staff of the Infantry and Cavalry School as a meritorious device, worthy of trial. The discussion filed with the invention pointed out, for the first time, the correct tactical employment of the weapon, and staked the military reputation and ability of the author and inventor on the correctness of his views.

From these facts it may be gathered that there was required a certain degree of originality and energy to get together and organize a machine gun battery for the Santiago campaign.

The project was conceived and executed. The service rendered by this battery has forever set at rest the question of the proper tactical use of the machine gun arm, both on the offensive and defensive. These things are now beyond the realm of theory. They are a demonstrated problem. The solution is universally acknowledged to be correct.

This is the history of that detachment.

CHAPTER 2
Inception

From the 26th of April until the 6th of June, Tampa and Port Tampa were the military centres of greatest interest in the United States. Troops were rushed into these places on special trains and camped on available sites, pending the organization of a proposed expedition to—somewhere. Supplies of every description came pouring in on long trains of express and freight cars; mounted officers and orderlies ploughed their rushing way through great heaps and dunes of ever-shifting sand, leaving behind them stifling clouds of scintillating particles, which filtered through every conceivable crevice and made the effort to breathe a suffocating nightmare. Over all the tumultuous scene a torrid sun beat down from a cloudless sky, while its scorching rays, reflected from the fierce sand under foot, produced a heat so intolerable that even the tropical vegetation looked withered and dying. In this climate officers and men, gathered mostly from Northern posts, were to "acclimate" themselves for a tropical campaign—somewhere.

They never encountered as deadly a heat, nor a more pernicious climate, in Cuba nor in Porto Rico, than that of southern Florida. Its first effect upon men just emerging from a bracing Northern winter was akin to prostration. Then began to follow a decided tendency to languor; after this one was liable to sudden attacks of bowel troubles. The deadly malaria began to insidiously prepare the way for a hospital cot; the patient lost flesh, relish of food became a reminiscence, and an hour's exertion in the sun was enough to put a man on his back for the rest of the day. Exposure to the direct action of the sun's rays was frequently followed by nausea, a slight chill, and then a high fever. The doctors subsequently called this "thermal fever," which is suspected to be a high-sounding name calculated to cover up a very dense ignorance of the nature of the disease, because no one ever obtained any relief from it from them. Recurrence of the exposure brought recurrence of the fever, and, if persisted in, finally produced a severe illness.

One reason for this was that the troops continued to wear the winter clothing they had worn on their arrival. The promised "khaki" did not materialize. Some regiments drew the brown canvas fatigue uniform, but the only use made of it was to put the white blanket-roll through the legs of the trousers, thereby adding to the weight of the roll, without perceptible benefit to the soldier.

Such a climate, under such surroundings, was not conducive to original thought, prolonged exertion, or sustained study. Everybody felt "mean" and was eager for a change. Nobody wanted to listen to any new schemes. The highest ambition seemed to be to get out of it to somewhere with just as little delay and exertion as possible. It was at this juncture that the plan of organizing a Gatling gun battery was conceived, and the attempt to obtain authority began.

The Gatling gun is one of the two machine guns adopted in the land service of the United States. Not to enter into a technical description, but merely to convey a general idea of its working and uses, it may be described as follows:

The gun is a cluster of rifle barrels, without stocks, arranged around a rod, and parallel to it. Each barrel has its own lock or bolt, and the whole cluster can be made to revolve by turning a crank. The bolts are all covered in a brass case at the breech, and the machine is loaded by means of a vertical groove in which cartridges are placed, twenty at a time, and from which they fall into the receivers one at a time. As the cluster of barrels revolves each one is fired at the lowest point, and reloaded as it completes the revolution. The gun is mounted on a Y-shaped trunnion; the lower end of the Y passes down into a socket in the axle. The gun is pointed by a lever just as one

points a garden hose or sprinkler, with the advantage that the gun can be clamped at any instant, and will then continue to sprinkle its drops of death over the same row of plants until the clamps are released. The axle is hollow and will hold about a thousand cartridges. It is horizontal, and on its ends are heavy Archibald wheels. There is also a heavy hollow trail, in which tools and additional ammunition can be stored. The limber resembles that used by the Artillery, and is capable of carrying about 9600 rounds of cartridges. The whole gun, thus mounted, can be drawn by two mules, and worked to good advantage by from six to eight men. It is built of various calibres, and can fire from 300 to 900 shots per minute. The guns used by the Gatling Gun Detachment, Fifth Army Corps, were built by the Colt's Arms Co., were the latest improved model, long ten-barrel gun, and fired the Krag-Jorgenson ammunition used by the Regular Army.

The attempt to obtain authority to organize a machine gun battery met with many discouragements and repeated failures. No one seemed to have thought anything about the subject, and Tampa was not a good place nor climate in which to indulge in that form of exercise, apparently. Perhaps the climate was one reason why so little thinking was done, and everything went "at sixes and sevens."

The officer who had conceived the scheme was a young man, too. He was only a second lieutenant ("Second lieutenants are fit for nothing except to take reveille"), and had never, so far as his military superiors knew, heard the whistle of a hostile bullet. He had made no brilliant record at the Academy, had never distinguished himself in the service, and was not anybody's "pet." He was, apparently, a safe man to ignore or snub if occasion or bad temper made it desirable to ignore or snub somebody, and, above all, had no political friends who would be offended thereby.

"Politics" cut quite a figure in Tampa in some respects. An officer who was known to be a personal friend of Senator Somebody, or *protégé* of this or that great man, was regarded with considerable awe and reverence by the common herd. It was ludicrous to see the weight attached to the crumbs of wisdom

Skirmish drill at Tampa

that fell from the friends of the friends of somebody. They shone only by a reflected light, it is true; but nobody there at Tampa had a lamp of his own, except the few who had won renown in the Civil War, and reflected light was better than none at all. A very young and green second lieutenant who was able to boast that he had declined to be a major in a certain State was at once an oracle to other lieutenants—and to some who were not lieutenants. The policy which governed these appointments was not so well understood at that date in the campaign as it is now.

When the court of a reigning favourite was established at the Tampa Bay Hotel as a brigadier, and people began to get themselves a little settled into the idea that they knew who was in command, they were suddenly disillusioned by the appointment of another and senior brigadier to the command. They settled down to get acquainted with the new authority, and were just beginning to find out who was who, when the telegraph flashed the news that the deposed potentate had been made a major-general, and, of course, was now in command. The thing was becoming interesting. Bets began to be made as to which would come in ahead under the wire. The other also became a major-general. Then came a period of uncertainty, because the question of rank hinged upon some obscure and musty record of forgotten service some thirty-four years before. From these facts will be apparent the difficulty under which a subordinate laboured in trying to create anything.

It is hardly worth while in any case of that sort to waste time with subordinates. The projector of an enterprise had better go straight to the one who has the necessary authority to order what is wanted; if access to him can be had, and he can be brought to recognize the merits of the plan—that settles it; if not—that also settles it. In either case the matter becomes a settled thing, and one knows what to depend upon.

But who was the man to see there at Tampa? Nobody knew.

The first officer approached was the one in direct line of superiority, Col. A. T. Smith, 13th Infantry. The idea was to ascertain his views and try to obtain from him a favourable endorsement upon a written plan to be submitted through military channels to the commanding general at Tampa. Perhaps it was the deadly climate; for the reply to a request for a few minutes' audience on the subject of machine guns was very gruff and curt: "I don't want to hear anything about it. I don't believe in it, and I don't feel like hearing it. If you want to see me about this subject, come to me in office hours." That settled it. Any effort to get a written plan through would have to carry the weight of official disapproval from the start, and even a "shavey" knows that disapproval at the start is enough to kill a paper in the official routine.

The next officers approached were Major William Auman and Capt. H. Cavanaugh, of the 13th Infantry, who were asked for advice. These two officers, both of whom rendered very distinguished services on the battle-field, listened with interest and were convinced. Their advice was: "Get your plan in tangible shape, typewritten, showing just what you propose; then go straight to the commanding general himself. If he listens to you, he will be the responsible party, and will have waived the informality; if he will not receive you, no harm is done."

This advice was followed and the following plan prepared:

SCHEME FOR ORGANIZATION OF DIVISION
GATING GUN DETACHMENT

MATERIAL
Three guns with limbers and caissons; 28 horses and 16 saddles; 6 sets double harness, wheel, and 6 lead; 1 escort wagon, team and driver; and 100,000 rounds, .30 cal.

PERSONNEL
One first lieutenant, 3 sergeants, 3 corporals, 1 clerk, 1 cook, and 35 enlisted men selected for their intelligence, activity, and daring; volunteers, if possible to be obtained, as the service will be hazardous.

EQUIPMENT
Officer: Revolver, sabre, or machete, and field-glass.
Enlisted men: Revolver and knife.

FIELD BAKERY

Fifty rounds to be carried on person for revolver, and 50 in ordnance train.

CAMP EQUIPAGE
Four conical wall-tents, 2 'A' wall-tents, and the ordinary cooking outfit for a company of 41 men.

ORGANIZATION
In the discretion of the detachment commander, subject to approval of division commander; probably as follows, subject to modifications by experience:
Three detachments under a sergeant. A detachment to be composed of 1 gunner and 7 men. The gunner should be a corporal.

ADMINISTRATION
The Division Gatling Gun Detachment to be subject only to the orders of the division commander, or higher authority. Its members are carried on 'd. s.' in their respective organizations. Its commander exercises over it the same authority as a company commander, and keeps the same records. Returns, reports, and other business are transacted as in company, except that the detachment commander reports directly to and receives orders directly from Division Headquarters. The detachment is not subject to ordinary guard or fatigue. When used as part of a guard, whole detachments go with their pieces.

INSTRUCTION
The organization is purely experimental; hence the greatest possible latitude must be allowed the detachment commander, and he should be held accountable for the results. He should not be subjected to the orders or interference of any subordinates, however able, who have made no special study of the tactical use or instruction for machine guns, and who may not have faith in the experiment. It will be useless to expect efficiency of the proposed organization unless this liberty be accorded its organizer. The field is a new one, not yet well discussed

by even the text-writers. Organization and instruction must be largely experimental, subject to change as the result of experience; but no change from the plans of the organizer should be made except for good and sufficient reasons.

TACTICAL EMPLOYMENT

This organization is expected to develop:
    (a) The fire-action of good infantry.
    (b) The mobility of cavalry.
Its qualities, therefore, must be rapidity and accuracy, both of fire and movement.

Its employment on the defensive is obvious. On the offensive it is expected to be useful with advance guards, rear guards, outposts, raids, and in battle. The last use, novel as it is, will be most important of all. The flanks of the division can be secured by this organization, relieving reserves of this duty; it will give a stiffening to the line of support, and at every opportune occasion will be pushed into action on the firing line. The *moral effect* of its presence will be very great; it will be able to render valuable assistance by its fire (over the charging line) in many cases. Last, but very important, the occupation of a captured line by this organization at once will supply a powerful, concentrated, and controlled fire, either to repulse a counter-charge or to fire on a discomfited, retiring enemy. Being a horsed organization, it can arrive at the critical point at the vital moment when, the defender's first line having been thrust out, our line being disorganized, a counter-charge by the enemy would be most effective, or controlled fire by our own troops on him would be most useful.

It is urged that this last use of machine guns is one of the most important functions, and one which has been overlooked by writers and tacticians.

There is one vital limitation upon the proposed organization; viz., it must not be pitted against artillery.

It is urgently suggested that this organization can be perfected here and now without difficulty, while it will be very difficult to perfect after the forward movement has begun. Horses and harness can be easily procured at Tampa; there will be no difficulty if some energetic officer be authorized to proceed with the work, and directed to attend to the details.

Believing earnestly in the utility of the proposed organization, which will convert useless impedimenta into a fourth arm, and realizing the dangerous nature of the proposed service, I respectfully offer my services to carry these plans into effect.

John H. Parker
2nd Lieut. 13th Infty.

With this plan well digested and with many a plausible argument in its favour all thought out, Col. Arthur McArthur, assistant adjutant-general to Gen. Wade, who was at that moment in command, was approached.

Col. McArthur was a very busy man. He was also a very business-like man, and one of handsome appearance, easy access, and pleasant address. He sandwiched in a fifteen-minute interview between two pressing engagements, and manifested both interest and approval. But nothing could be done at that time. "Come again a week from to-day," said he, "and I will try to obtain you a hearing before one who can do what you wish by a single word. I believe in your scheme and will help you if I can." The week rolled by and a change of commanding generals occurred. Gen. Wade was ordered away, taking McArthur with him, and no progress had been made. It was discouraging.

The next step in the plan was by lucky accident. Lieutenant (now Lieut.-Col.) John T. Thompson, Ordnance Department, who was in charge of the Ordnance Depot at Tampa, accidentally met the would-be machine-gun man, and was promptly buttonholed over a dish of ice cream. Thompson was himself a young man and a student. His department placed an insuperable obstacle in the way of himself carrying out a plan which he, also, had conceived, and he was keen to see the idea, which he fully believed in, demonstrated on the battle-field. He had, moreover, as ordnance officer, just received an invoice of fifteen

AWAITING TURN TO EMBARK

Gatling guns, complete, of the latest model, and he had access to the commanding general by virtue of being a member of his staff. By reason of the terrible rush of overwork, he needed an assistant, and it seemed practicable to try to kill two birds with one stone. But all he said was, "I believe in the idea; I have long advocated it. It may be possible for me to get you your opportunity, and it may not. If so, you will hear from the matter."

The attempt to get the thing going had been apparently abandoned, when, utterly without notice, the regimental commander received orders per letter, from Headquarters Fifth Army Corps, which resulted in the following orders:

Headquarters 13th Infantry
In the Field
Tampa, Fla.
May 27, 1898

SPECIAL ORDERS No. 22:

Pursuant to instructions contained in letter from Headquarters 5th Army Corps, May 26, 1898:

2nd Lieut. John H. Parker, 13th Infantry
Sergeant Alois Weischaar, Company A
Sergeant William Eyder, Company G
Private Lewis Kastner, Company A
Private Joe Seman, Company B
Private Abram Greenberg, Company C
Private Joseph Hoft, Company D
Private O'Connor L. Jones, Company D
Private Louis Misiak, Company E
Private George C. Murray, Company F
Private John Bremer, Company G
Private Fred H. Chase, Company H
Private Martin Pyne, Company H,

will report to Lieut. J. T. Thompson, ordnance officer, for duty in connection with the Gatling Gun Battery.

These men will be fully equipped, with the exception of rifle, bayonet, scabbard, and blanket-bag, and will be rationed to include May 31, 1898.

By order of Colonel Smith.

M. McFarland

1st Lieut. 13th Infty., Adjutant.

These men were selected by their company commanders. It is not known whether the selections were made with a view to special fitness or not. They had no notice that the detail was to be anything but a transient character; in fact, one company commander actually detailed the cook of his private mess, and was intensely disgusted when he found that the detail was to be permanent or semi-permanent. The men were sent fully armed and equipped; carrying rifles, knapsacks, etc., and marched down to the Ordnance Depot for instructions. These instructions were to return to camp, turn in their rifles, bayonets, cartridges, belts, and knapsacks, and return early the following morning equipped with blanket-roll complete, haversack, and canteen. Each man, after full explanation of the hazardous duty, was given a chance to withdraw, but all volunteered to stay.

The instructions were obeyed, and the Gatling Gun Detachment was born—a pygmy.

CHAPTER 3
The Ordnance Depot

The Ordnance Depot at Tampa was located on Lafayette Street, at the end of the bridge over the river, next to the Tampa Bay Hotel. The river washed the sides of the building, which was occupied by the Tampa Athletic Club, and had formerly been used as a club-house. There were two stories and a basement. The basement was nearly on a level with the river, the main floor on a level with the bridge, and there was also a spacious upper floor. The main floor was used for storage of light articles of ordnance; the basement for heavy articles and ammunition. Hundreds of thousands of rounds of rifle and revolver ball cartridges, thousands of rounds of Hotchkiss fixed ammunition, and many hundreds of pounds of powder charges for field artillery and mortars were here stored. Miscellaneous assortments were daily coming in, generally without any mark on the box by which to learn what were the contents. The name of the arsenal, if from an arsenal, was usually stamped on the seal; generally there was no mark whatever to designate the origin or contents of the many boxes which came from ordinary posts. The invoices came from a week to ten days behind or in advance of the arrival of the boxes, and there was not the slightest clue to be gained from them. Consequently those who had to check up invoices and prepare for issues were at their wits' end to keep things straight. A requisition for so many articles would come in, duly approved; unless the boxes containing these articles happened to have been unpacked, it was uncertain whether they were on hand or not. No wholesale merchant of any sense would ship out boxes of goods without some indication of their contents; but that was exactly what was done from all over the country to the Ordnance Depot at Tampa.

The upper floor consisted of one large room. A rope railing was placed around it to preserve clear space around the desks. There were several of these for the ordnance officer and the various clerks. A chief clerk, an assistant clerk, a stenographer, and two ordnance sergeants looked after the red tape. An overseer with four subordinates and a gang of negro stevedores attended to loading and unloading boxes, storing them, counting out articles for issue or receipt, and such other duties as they were called on to perform. There was an old janitor named McGee, a veteran of the Civil War, whose business it was to look after the sweeping and keep the floors clean.

Four guns in their original boxes were issued to the detachment on the 27th of May. They were new, and apparently had never been assembled. On assembling them it was found that the parts had been constructed with such "scientific" accuracy that the use of a mallet was necessary. The binder-box on the pointing lever was so tight that in attempting to depress the muzzle of the gun it was possible to lift the trail off the ground before the binder-box would slide on the lever. The axis-pin had to be driven in and out with an axe, using a block of wood, of course, to prevent battering. A truly pretty state of affairs for a gun the value of which depends on the ease with which it can be pointed in any direction.

Inquiry after the war at the factory where the guns are made disclosed the fact that these parts are rigidly tested by a gauge by the Government inspectors, and that looseness is regarded as a fatal defect. Even play of half a hundredth of an inch is enough to insure the rejection of a piece. The very first thing done by the Gatling Gun Detachment, upon assembling these guns, was to obtain a set of armourers' tools and to file away these parts by hand until the aim of the piece could be changed by the touch of a feather. The detachment was ordered to rely upon the friction clutches for steadiness of aim, when necessary, and not upon the tight fit of the parts. It was ordered that there must be no doubt whatever of easy, perfectly free manipulation at any

BAIQUIRI

and all times, even if the pointing lever should become rusted. This precaution proved on July 1st to have been of great value.

The instruction of the detachment began immediately, and consisted, at first, of unpacking, mounting, dismounting, and repacking the guns. The four guns were mounted and a drill held each time in the loading and firing of the piece. This system of instruction was continued until the detachment was ordered on board ship on the 6th of June. During this instruction members of the detachment were designated by name to fall out, and the remainder of the detachment required to execute all the manoeuvres of the piece as before. In fact, this instruction was carried to such a point that one man alone was required to load, aim, and fire the gun at designated objects without any assistance.

The detachment at once assumed the position of an independent command. It reported directly to Maj.-Gen. W. R. Shafter, commanding the 5th Corps, in everything so far as its duties with Gatling guns were concerned, was regarded as an independent command, kept its own records in the same manner as a company, obtained cooking utensils from the quartermaster and ran its own mess, and furnished its own guard. This status, that of a separate command, continued until the detachment was finally disbanded at Montauk.

On the 27th of May the detachment commander was summoned to Gen. Wheeler's headquarters and there requested to explain to the general in person his plans for organizing a Gatling gun detachment. Gen. Wheeler had just assumed command of all the Cavalry belonging to the 5th Army Corps. His headquarters, instead of being in a suite of rooms in the palatial Tampa Bay Hotel, where all the other general officers had their headquarters, were located about half a mile from the hotel in a treeless pasture. The cavalry guidon floating from a lance-head was the only indication of headquarters, and the half-dozen "A" tents in an irregular line gave no sign that one of the most distinguished generals in the world had here his headquarters in the field.

The general was easily accessible. The first thing that impressed one of him was his extraordinary quickness. His eye seemed to take in everything within sight of him at a single glance, and to read one's thoughts before the tongue could give expression to them. He grasped ideas when they were only half uttered and immediately drew deductions from mere statements of simple facts, the result of years of careful study. These deductions, which Gen. Wheeler drew instantly, were in every case correct, and showed a keener and more correct appreciation of the proper tactical employment of machine guns than was shown by any other officer of the 5th Corps. The result of the interview with the general was that a scheme for the organization of a tactical unit to be composed of three Gatling guns and to be employed with the cavalry division, was drawn up on the spot, under Gen. Wheeler's personal direction, and was submitted by him to Gen. Shafter, with the request that authority be granted for the organization of this command for the purpose indicated.

In the application Gen. Wheeler stated that he believed that such a battery of machine guns, if properly handled, could go anywhere that cavalry could go, could take the place of infantry supports, could dash up and hold any ground or advantageous position that a body of cavalry might seize, could be thrown out to one flank of the enemy and assist in his demoralization in preparation for the cavalry charge, and would be of particular service in case the enemy attempted to form infantry squares, which were at that time supposed to be the main part of the Spanish tactics of battle. This application was disapproved.

On the 30th of May, Gen. Lee sent for the detachment commander for an interview on the subject of Gatling guns. Gen. Lee was at this time quartered at the Tampa Bay Hotel, and was engaged in the organization of the 7th Army Corps. It was supposed that the 7th Corps was designed for the Havana campaign, and it was believed that the attack upon Havana would begin at a very early date. The result of the interview with Gen. Lee was that he directed a scheme for the organization of a tactical unit to be composed of 9 guns, 3 batteries of 3 guns each, to be prepared for service with the 7th Army Corps.

THE *HORNET*

It was desired that this organization be a volunteer organization, and the application was therefore made for authority from the President, under that law of Congress authorizing the employment of special troops. Col. Guild, well and favourably known from his connection with the Massachusetts National Guard, was prepared to furnish a volunteer organization already in existence, well drilled and already officered, composed of the flower of the youth of Massachusetts, very largely of college graduates, who had already been communicated with on the subject, and who were even at that time expecting momentarily a telegram calling them to this duty. Nothing resulted from this effort.

Meantime the drill instruction of the little detachment continued. Its members had acquired a considerable degree of proficiency in the mechanical handling of their guns, and were beginning to appreciate the destructive possibilities of their weapon. They were enjoying a degree of liberty which they had not found in their regimental camp, because when not on duty they were free to come and go at will, when and where they pleased. The hours for instruction were designated in the morning and in the cool of the afternoon, leaving the middle of the day and the evening for the men's own recreation. The result of this system of treatment was that *esprit-de-corps* began to be developed in the detachment. They began to feel that they were a special organization, expected to do special work, and that they were receiving very special treatment. They began to be proud of being members of the Gatling Gun Detachment, to take greater interest in the work, and when on the first of June they received their monthly pay not a single member of the detachment committed any excesses in consequence of this unusual degree of freedom. No one was intoxicated. No one was absent without permission.

The detachment had not been at the Ordnance Depot very long before an opportunity occurred for some of its members to exhibit those qualities which made the success of the battery so conspicuous on the battle-field afterward. The detachment commander had been detailed by verbal orders on the first of June in charge of the issues of ordnance property to the Santiago expedition. This was in addition to his duties with the Gatling guns. The work would commence about 6 o'clock in the morning, and from that time until dark there was a continual stream of wagons carrying away stores such as rifles, haversacks, meat ration cans, tin cups, and all the articles needed by troops in the field during a campaign. The ammunition which was issued to the troops at this time was drawn at the same place.

When wagons arrived to receive issues, stevedores were directed to count out the different articles under the direction of an overseer, and these piles of articles were verified by the officer in charge of the issues. The stevedores then loaded them on the wagons which were to haul them to the different camps. Receipts in duplicate were always taken and invoices in duplicate were always given, in the name, of course, of Lieut. John T. Thompson, who was responsible for the stores.

On the 4th of June issues were being made of rifle-ball cartridges. These cartridges came packed in boxes of 1000 rounds each, and each box weighed 78 pounds. A great quantity of it was stored in the basement, where there was also a considerable quantity of fixed Hotchkiss ammunition, as well as several thousand rounds of powder charges in boxes. The Hotchkiss ammunition, which comes with projectile and powder both set in a brass case, is bad ammunition to pack; for, no matter how carefully it is handled, there is almost always some leakage of powder from the cartridge case, thus causing a certain amount of loose powder to sift into the box in which it is packed.

About half past 11 o'clock on this morning a negro stevedore accidentally dropped a box of rifle ammunition near a pile of Hotchkiss fixed, and the next instant the labourers saw smoke ascending toward the ceiling of the basement. They yelled "Fire! fire!" at the top of their voices, and everybody in the basement at once made a rush for the two doors. It was a panic. The danger was imminent. The smoke curled up to the ceiling and then curled down again, and the excited, panic-stricken faces of the negroes as they rushed through the door made an awful picture of human terror. People on the outside of the building began to shout "Fire!"

At this juncture McGee, the old janitor, who had just reached the door, cried out, "Lieutenant, there is a box in here on fire!" speaking to Lieut. Parker, who was verifying issues just outside the door.

The lieutenant replied, "Let's throw it into the river," and dashed toward the box through the door, pushing the excited negroes to each side in order to assist McGee, who had instantly started for the box.

When Lieut. Parker reached the box, he found that McGee had already taken it up, and was staggering under its weight. He placed one arm around McGee's shoulder and with the other assisted him to support the box, from which the smoke was still ascending, and the two rushed for the door, throwing the whole momentum of their weight and speed against the crowd of frightened negroes, who were falling over each other in their panic-stricken efforts to escape. Priv. Greenberg, of the 13th Infantry, a member of the Gatling Gun Detachment, who was the sentinel on post at the time, saw the two men coming with the box, and with great presence of mind added his own weight with a rapid rush to the shock they had produced, thus enabling them to break their way through the dense throng at the door. It was only the work of an instant to then throw the box in the river, where it sank in the water and for a moment the blue smoke continued to bubble up from the box, which lay clearly visible on the bed of the river, the water being only about two feet deep at this point, which was, however, enough to entirely cover the box and thus extinguish the fire. At the outcry of "Fire!" Lieut. H. L. Kinnison, of the 25th Infantry, who was waiting outside of the basement with a wagon, started in at the other door, and Serg. Weischaar, acting first sergeant of the Gatling Gun Detachment, started for water. Just as the two men emerged from the door carrying the box, Lieut. Kinnison reached the spot where the fire had originated, and Serg. Weischaar appeared with two buckets of water. He and Lieut. Kinnison at once flooded the floor, seized a woollen cloth which happened to be near, and wetted down the boxes of Hotchkiss ammunition as a measure of precaution.

McGee, the hero of this episode, is an old veteran of the Civil War, having served three years in the Pennsylvania Volunteer Infantry during the war, and five years in the Regular Army after the war. He has never drawn a pension nor applied for one, although he suffers considerably from disease and wounds contracted and received during the war, and certainly should be rewarded by a grateful government for his conspicuous heroism. The explosion of this magazine would have brought the whole expedition to a standstill, besides inflicting tremendous destruction of property and frightful loss of life.

The same day the Artillery of the army began to draw its material for the campaign, and for a period of thirty-nine hours there was no rest for anybody connected with the issue of ordnance stores. It was at this time that the lack of intelligent marking and packing of the boxes was keenly felt. The greatest difficulty was experienced in selecting, from the mass of stores in the depot, the stores that were required by the Artillery. It was especially difficult during the work by night, when the only light that could possibly be allowed was a single lantern, on account of the danger of fire.

At the close of this thirty-nine hours of arduous duty, the officer in command of the Gatling Gun Detachment learned that orders had been issued for the embarkation of the 5th Army Corps at Port Tampa, and that no reference had been made to the Gatling Gun Detachment in these orders. He at once sought Lieut. Thompson, who could offer no light on the omission, but said:

"Have orders to send at once to the *Cherokee* 521,000 rounds of rifle-ball cartridges and all the revolver ammunition on hand. This is the reserve ammunition of the 5th Army Corps. I will send you in charge of this ammunition and you will see it to its destination. You may take an escort or not, as you please. The ammunition is to go on the 4 o'clock train and you must make all the arrangements in regard to it. Get box-cars, haul the ammunition over there and put it in the cars, see that it goes on

that train, and as soon as it arrives at Port Tampa, see that it is properly put on board the *Cherokee*."

In order to fully understand the situation of the Gatling Gun Detachment at this juncture, the following correspondence on the subject is necessary:

Office of Ordnance Officer
Lafayette Street
West of Bridge
Tampa, Fla.
June 3, 1898

The Assistant Adjutant-General
5th Army Corps, Tampa, Florida:

Sir,

Replying to your letter of June 1, 1898, in reference to Gatling Gun Detachment, I have the honour to submit the following report:

Guns, men, and equipment required for a 4-gun detachment:

	Guns	Serg	Corp	Priv
Total Required	4	5	4	28
On Hand	4	2	0	10
Required		3	4	18

The gun crews thus organized will give most effective service for the detachment.

Ammunition: Each limber carries 9,840 rounds cal. .30. Four limbers, 27,360; necessary reserve, 32,640; total, 60,000.

Tentage: Two conical wall-tents for enlisted men; one 'A' wall-tent for officer.

Camp equipage, in addition to that on hand in Gatling Gun Detachment: one *buzzacot*, small; four mess-pans, one dish-pan, one coffee-mill.

Blanket-roll complete; revolver with 50 rounds per man; waist-belts and entrenching-knives.

It is recommended that Priv. Butz, 'G' Co., 13th Infantry,

Corp. Robert S. Smith, 'C' Co., 13th Infantry, and Serg. Weigle, 9th Infantry, be members of the detachment; and that detachment be taken from 9th Infantry, which has some well-instructed men.

It is further recommended that the detachment be fully horsed as soon as practicable, and that the whole be placed under the command of Lieut. John H. Parker, 13th Infantry, as acting captain.

I recommend that I be authorized to issue the 4 Gatling guns and parts to him.

The details should carry the rations prescribed in General Orders 5th, May 31, 1898, 5th Army Corps. Very respectfully,

Jno. T. Thompson
1st Lieut., Ord. Dept, U. S. A.

This letter, prepared by Lieut. Parker and signed by Lieut. Thompson, was endorsed as follows:

First Endorsement:

Headquarters 5th Army Corps
Tampa, Fla.
June 5, 1898

Respectfully returned to Lieut. J. T. Thompson, Ordnance Officer.

If Lieut. Parker, in charge of the detachment as at present constituted, can make the arrangements suggested within, he may take action; but, in view of the limited time remaining, it is thought the detachment already organized will answer.

By command of Maj.-Gen. Shafter.

E. J. McClernand
Assistant Adjutant-General

Second Endorsement:

Office of the Ordnance Officer
Lafayette Street Bridge
Tampa, Fla.

WRECKED LOCOMOTIVES AND MACHINE SHOPS AT BAIQUIRI

June 5, 1898
Respectfully referred to Lieut. John H. Parker for his information.
Jno. T. Thompson
1st Lieut., Ordnance Dept, U. S. A.

It will be seen from the first endorsement that a certain amount of discretion was left to the detachment commander. He was authorized to take action if he could make the arrangements suggested within. Lieut. Thompson had authorized an escort for the reserve ammunition, if it was considered necessary. The detachment commander resolved to take action by using his whole detachment as an escort, putting it on board the *Cherokee*, with the reserve ammunition, and accompanying it to its destination—in Cuba, trusting to the future to enable him to complete the detachment according to the first endorsement.

It was now 11 o'clock in the forenoon. Between that time and 4 o'clock it was necessary to obtain two freight cars, have them placed upon the siding at a convenient point, have more than twenty wagon-loads of ammunition, camp equipage, etc., placed in these cars, have the four guns with their limbers placed on board, and, more difficult than all the rest, go through the necessary red tape at the quartermaster's office in order to get the two cars moved to Port Tampa. It was all accomplished.

The general freight agent was bluffed into believing that unless the two cars were instantly set where they were wanted his whole railroad would be tied up. The quartermaster was hypnotized and dropped formality, putting all the clerks to work upon papers and making out the necessary bill of lading, invoices, etc., in time to catch the 4 o'clock train. He also issued the necessary transportation for the officer and men of the detachment from Tampa to Port Tampa, accepting the first endorsement above as sufficient orders for that purpose.

One member of the detachment, Priv. Murray, had been very ill with what we afterward learned to call the Cuban fever, and, while apparently convalescent, was entirely too weak to accompany the detachment. He was a splendid fellow, and the tears rolled down his emaciated face when he was told he must remain behind. He was furnished with a descriptive list and a letter was written to the chief surgeon of the Division Hospital, requesting him to send an ambulance immediately for the sick man. One member of the detachment carried this letter to Tampa Heights, and so sharp was the work of getting away that this man had to board a moving train as it was pulling out to keep from getting left; but Priv. Murray was taken to the hospital and cared for, and Priv. Bremer did not get left.

The detachment reached Port Tampa about sundown, and Maj. Cushing, who had charge of the loading of the transports, at once authorized the cars to be set alongside the *Cherokee*. The ammunition, guns, camp equipage, men, and all were promptly put aboard. The training in packing and unpacking the guns was the only thing which enabled the work to be done in the limited time allotted. Not so much as a ten-penny nail belonging to the detachment was left behind.

During the night the troops that were to occupy the *Cherokee* came on board, and it was found the next morning that five or six tons of regimental baggage had been piled on top of the guns, making it practically impossible to disembark, even if such a movement should be ordered.

CHAPTER 4

The Voyage and Disembarkation

It seemed that the work had been accomplished none too soon, for on the morning of June 7th orders came to the *Cherokee* to leave the slip and proceed down the bay. There were on board at this time, beside the little Gatling Gun Detachment, the 17th Infantry, under command of Col. Haskell, and a battalion of the 12th Infantry, under command of Col. Comba, who was the senior officer on board. The ship was frightfully crowded. The berth deck and lower deck had been arranged for the accommodation of the men by nailing rows of two 2x4 scantlings just far enough apart to leave room for a man to lie down, and fastening three tiers of bunks to these scantlings. The men were packed in these bunks like sardines in a box. The ventilation was conspicuous by its absence, the heat below deck was frightful and the misery entailed by such accommodations was beyond description. But the men were very cheerful, and, being allowed the privilege of the upper deck, very little in the way of complaint was heard. Everybody was anxious to be off. The hope most frequently expressed was for a quick passage and a sharp, swift campaign. It was easily foreseen by the officers on board the ship that a long sojourn on shipboard under such conditions would have a very bad effect on the men.

The ship dropped down the bay to the quarantine station, starting about noon, and there lay to, waiting, as was supposed, for the remainder of the fleet. Suddenly, about 8 p. m., one of the torpedo cruisers came tearing down the bay under full steam, and we heard the message sounded through the megaphone: "Return to port. Three Spanish cruisers within three hours' sail of the offing." It was a thrilling moment. Officers and men were lounging, taking, as they supposed, their last view of the American shores, without a suspicion of present danger, when they were rapidly brought to a realizing sense that "war is hell," by a notice that the enemy was upon them. Whether they were in danger or not, the danger was deadly real and imminent to them at the time.

The *Cherokee* had been anchored pretty well inside. She immediately got up steam and went out to warn other vessels farther out in the offing, and then made safely for the harbour. Officers and men behaved with perfect coolness. It was hopeless to attempt to escape by concealment, so Col. Comba ordered out the band of the 17th Infantry and the good ship fled up the bay, in momentary expectation of a smashing shot from the enemy, to the strains of "There'll be a hot time." What little excitement there was displayed itself in a feverish searching of the bay with field-glasses for signs of the enemy. The older officers, upon whom the responsibility was resting, sat upon the quarter-deck, smoking their pipes and discussing the situation. The captains quietly moved about, assigning stations to their companies, in case of attack, with the view of trying the effect of the modern rifle upon the armoured sides of a Spanish man-of-war, and two of the younger officers took advantage of the catchy air which the band was playing to dance a two-step on the quarter-deck. So the evening wore away. The moon went down. The myriad little stars came out, twinkling in the deep blue sky, and at last both officers and men, tired of looking for an enemy who was never to appear, turned in for such sleep as they could get, leaving a small guard on deck to keep a lookout. When they awoke next morning, the ship was in the deepest part of the nearest slip, moored fast by her guy-ropes to the dock. Thus ended the first engagement with the enemy.

From the 8th until the 13th, the *Cherokee* lay at anchor in the slip. She was relieved on the 10th of about 200 men, thus slightly lightening her overcrowded condition. In the meantime, this overcrowded condition of the ship had led to some discussion as to who could best be moved on board some oth-

THE LANDING

er ship, with some prospect that the Gatling Gun Detachment might be disturbed. The situation was not at all satisfactory. With four guns, no mules, no harness, no authority, and only twelve men, the Gatling Gun Detachment did not appear to be in a very fair way toward inflicting much damage upon the enemy. So on the 11th of June the detachment commander visited Gen. Shafter at his headquarters, determined to bring the matter to an issue, definitely, one way or the other. This was the first time he had met the general, and, under the circumstances, the manner of his reception appeared to be doubtful.

Gen. Shafter is a big man. This is not noticed at first glance. He is above the average height, but his corpulent figure does not indicate that he is full five feet nine inches in height, because his girth is of like proportion. His hands are big; his arm is big; his head is big. The *occiput* is especially full, and the width of head just over the ears is noticeable. There is plenty of room for the organs of combativeness. One would think he is probably a lover of children; during this interview he patted the head of an inquisitive dog, which evidently belonged somewhere on board the flag-ship, and which strayed into the room. His eyes are big, very full and very keen. As you enter he says curtly, "Take a seat." He waits, looking down, for you to state your business, then suddenly fixes you with a piercing glance, and goes to the heart of the subject by one incisive sentence, which leaves no more to be said. This description is a general type of several interviews with him. On this occasion the general inquired concerning the facts, looking keenly, searchingly, and meditatively at the detachment commander. The machine gun man was "on trial."

Then the general broke the silence by one short question, "What do you want?"

And the reply was in kind, "Twenty men, general, with the privilege of selecting them."

The general suggested the advisability of taking a complete organization; to which was replied: "That at this late hour in the expedition it is imperative to have selected men in order to perform the required duty; that men taken at random, as would be the case in a complete organization such as a company, would not be likely to have the required characteristics."

The general tersely remarked, "You may have them. Make out your list, name any man in the corps that you want, and hand the list to me. I will send the men to you."

The trial was over, and the Machine Gun Detachment was a settled fact.

Accordingly on the following day Special Orders No. 16 were issued, extracts follow:

Headquarters 5th Army Corps
On Board S. S. *Seguransa*
Tampa Bay, Fla.
June 11, 1898

SPECIAL ORDERS, No. 16

4. The following named enlisted men are detailed for duty with the Gatling Gun Detachment, 5th Army Corps, and will report at once to 2nd Lieut. John H. Parker, 13th Infantry, commanding the detachment for duty:

9th Infantry: Sergeant Weigle.

12th Infantry: Privates Voelker, Company A; Anderson, Lauer, and Timberly, Company C; Prazak, Company E.

13th Infantry: Sergeant Green, Company H; Corporals Stiegerwald, Company A; Doyle, Smith, and Rose, Company C; Privates Corey and Power, Company A; Barts, Company E; and Schmadt, Company G.

17th Infantry: Privates Merryman and Schulze, Company A; McDonald, Company B; Elkins, Dellett, and McGoin, Company D; Click, Needle, Shiffer, and Sine, Company E.

Each of the soldiers will report equipped as follows: Blanket-roll complete, haversack and contents, canteen, waist-belt of leather, hunting-knife, and revolver, and they will be rationed with ten days' travel rations. Descriptive lists of these men will be sent to the commanding officer of the detachment.

By command of Maj.-Gen. Shafter.

Official. *J. D. Miley*

E. J. McClernand, Aide. Asst. Adj.-Gen.

PACK TRAIN

Headquarters 5th Army Corps
On Board S. S. *Seguransa*
Tampa Bay
June 11, 1898

SPECIAL ORDERS, No. 16

5. 2nd Lieut. John H. Parker, 13th Infantry, commanding the Gatling Gun Detachment, 5th Army Corps, is authorized to make the usual requisitions for supplies.
By command of Maj.-Gen. Shafter.
Official. *J. D. Miley*
E. J. McClernand, Aide. Asst. Adj.-Gen.

The organization was thus perfected by a single stroke of the general's pen on the 11th of June, theoretically; practically it was the 14th of June before the details from the 12th and 17th Infantry reported, and when they did, instead of being equipped as directed, they carried rifles with 100 rounds of ammunition.

Serg. Weigle, of the 9th Infantry, who reported at the same time, carried a revolver. On the 14th a wigwag message was received from the 13th Infantry, inquiring whether the detail was desired to report at once or not, to which the reply was sent that it was desired to report at the earliest possible moment. It did not report.

The detachment was at once organized as well as possible for the trip on board the transport, and the guns brought up from the hold of the ship and mounted in such a way that they would be ready for instant use. It was not known but that the detachment might have to participate in a naval engagement, and the value of machine guns in the navy has long been demonstrated. At any rate, it was determined to be ready to give a warm reception to any torpedo vessel which might attempt to attack the *Cherokee*. One object of getting the guns up was to give instruction to the new men who reported on the 14th. Sergt. Weigle was well instructed in the use of Gatling guns, but none of the other members of the detachment had ever received any instruction, and had been selected rather on the ground of their superior intelligence and courage than on any special knowledge of machine guns. They were given a drill each day in loading and firing the piece, during the time they remained on board the transport, when the weather permitted.

The condition of the troops on board the transport was miserable. The following extract from a letter written at that time will convey some idea of the crowded, ill-ventilated condition of the vessel:

> We have now been on board the transport a week, and are getting into a frame of mind suitable for desperate work. If you can imagine 1000 men crowded into space needed for 500, and then kept there without room to stand or move or sit for seven days, under a tropical sun, in foul holds utterly without ventilation (just imagine it!), endured without a single murmur or complaint, not stoically, but patiently and intelligently, while every officer on board is kicking as hard and as often as possible for the relief of his men, then you will have some idea of the situation. The men are very patient, but they know someone has blundered. Talk about the heroism of the Light Brigade! It is nothing to the heroism that goes cheerfully and uncomplainingly into the Black Hole of Calcutta (there is nothing else that will compare with these transports), all because it is duty. When will the people appreciate the heroism of the Regular Army?

This was the actual condition of affairs on board the *Cherokee* up to the time of leaving port on the 14th of June, and it was modified only by the hoisting of wind-sails, after we got under way. These were not very efficient and there were only two of them, so very little relief was given to the overcrowded berth-deck. Most of the men spent their time on the upper deck, and one whole company was quartered there. At night, after 8 o'clock, Col. Comba authorized the men to sleep on deck, and there was always a rush, when the ship's bell struck the hour, for good places on the quarter-deck. The only thing that made the voyage endurable was the good weather which prevailed. This prevented seasickness, to a certain extent.

The squadron reached Santiago de Cuba, and after tacking about for several days, either for the purpose of deceiving the enemy, or of waiting a decision as to the landing-place, finally approached Baiquiri, which had been selected for the landing. The troops on the *Cherokee* began to land on the 23rd of June, the battalion of the 12th Infantry going first. This was followed by the 17th Infantry, and upon its departure the captain of the *Cherokee* put to sea. The reason for this manoeuvre is not known. The orders issued by Gen. Shafter in regard to the landing were that the Gatling Gun Detachment should accompany Gen. Lawton's Division. This movement of the *Cherokee* completely blocked the landing of the Gatling guns. The ship's captain was finally induced to put back into the bay and speak to the *Seguransa*, and Gen. Shafter directed that the detachment should be taken off the next morning.

An effort was made, therefore, to obtain the use of a lighter which was not at that time in use, but the Commissary Department refused to yield the boat, and it remained until 11 o'clock the next morning tied up to the wharf with half a load of commissaries on board before it became available, and then was seized by the Quartermaster's Department. An effort was then made to obtain the use of three pontoons, belonging to the Engineer Department, which had been drawn up to the shore and were of no use to anybody. The young engineer officer in charge of these boats, a premature graduate of the class of '98, was "afraid the boats might get smashed in the surf," and could not consent without seeing Col. Derby. Col. Derby could not be found.

A wigwag came from Gen. Shafter, asking whether the Gatling guns had been landed. The reply, "No; may I use pontoons?" was answered at once, "Use pontoons, and get off immediately."

On returning to shore with a party to work the pontoons, the party was stopped in the act of launching the first boat by Gen. Sumner, and ordered to proceed to the *Cherokee*, take her out into the offing, and order another to take her place to unload. Protesting against this action, and informing Gen. Sumner of the urgent orders for the Gatling guns to disembark at once, that officer inquired the opinion of the prematurely graduated engineer as to the practicability of using the pontoons, and this experienced young man again expressed the fear that the boats might be injured in the surf. To the detachment commander's indignant exclamation, "What the h—were these boats made for, if they are not to be used and smashed?"

Gen. Sumner responded by a peremptory order to warp the *Cherokee* out from the pier and send the other vessels in. The order was obeyed, and all the circumstances reported to Gen. Shafter the same evening, with the expression of the opinion that if the general wanted the Gatling guns landed, he would have to attend to it personally, because the Gatling gun commander did not have sufficient rank to accomplish it in the face of all these obstacles.

Early on the morning of June 25th, therefore, Gen. Shafter sent peremptory orders to the lighter to lay alongside the *Cherokee*, take the Gatling guns and detachment on board, and land them on the dock. The transfer began at 8 o'clock in the morning, Gen. Shafter coming out in person in his steam launch to see that his order was executed.

By 11 o'clock the guns, carriages, 30,000 rounds of ammunition, four sets of double harness, and the detachment were on board the lighter. This had been accomplished a mile outside in the offing, with the vessel rolling and pitching in the trough of the sea and on the crest of the gigantic rollers in so violent a manner that it was almost impossible for men to stand on their feet, much less handle such heavy material as guns and ammunition.

The lighter was warped to the pier at 11 o'clock, and the general tied his steam launch alongside to see that it was not disturbed until the debarkation was completed. At 1 o'clock everything was ashore, and, in compliance with the general's instructions, the best mules in the corral were taken, and as they were led away from the corral-gate, a fat, sleek, black streaked, long-eared specimen, which had been selected for a saddle-mule, set up a cheerful *"Aw! hee haw! haw!"* which produced a burst of laughter and cheering from the members of the detachment and the soldiers in the vicinity. It was a cheerful omen. These

Missouri mules were capable of pulling anything loose at both ends, and four experienced drivers had been selected from the detachment who were capable of riding anything that walked on four feet, or driving anything from an Arab courser to a pair of Shetland ponies.

Priv. J. Shiffer had been selected as corral boss of the detachment. The most picturesque figure, the most boyish member, and as brave a soldier as ever shouldered a musket; broad of shoulder, stout of limb, full of joke, as cheerful as a ray of sunlight, this man was the incarnation of courage and devotion. He loved a mule. He was proud of the job. With the instinct of a true teamster, he had snapped up the best pair of mules in the whole corral and was out before the detachment commander had selected a single mule. This team was as black as Shiffer's shoes and as strong as a pair of elephants. They were worked harder than any other team in the 5th Army Corps, and when they were turned in to the quartermaster in August, they were as fat, as sleek, as strong, and as hardy as on the day they were taken from the corral in Baiquiri.

The other three teamsters were like unto the first. They were all handy men. They were as capable of fighting or aiming a gun as of driving a team. Any one of the four could take a team of mules up a mountain-side or down a vertical precipice in perfect safety. They could do the impossible with a team of mules, and they had to do it before the detachment reached the firing-line. The success of the battery was to depend to a very large degree upon the coolness, good judgment, and perfect bravery of these four teamsters.

It should be noted that the use of mules was an experiment. The "scientific" branch of service has always held that the proper animal to draw a field-piece is the horse. They expatiate with great delight upon the almost human intelligence and sagacity of that noble animal; upon his courage "when he snuffeth the battle afar," and upon the undaunted spirit with which he rushes upon the enemy, and assists his master to work the destruction of his foes. The Artillery claims that mules are entirely too stubborn, too cowardly, and too hard to manage

CAVALRY PICKET LINE

for the purpose of their arm of the service. It was also an experiment to use two mules per gun.

The Engineer Department had reported that the road to the front was impassable for wheeled vehicles, and even the general had apparently thought that four mules per gun would be necessary. The necessity of economizing mules, and the opinion of the detachment commander that two mules per gun would be sufficient, had led to the issue of that number. Those who despise the army mule for the purposes of field artillery know very little of the capacity of this equine product of Missouri when properly handled. It was demonstrated that two mules can pull a Gatling gun with 10,000 rounds of ammunition, loaded down with rations and forage, where eight horses are required to draw a field-piece; and that mules are equally as easy to manage under fire as horses.

The landing was completed and the detachment organized at 3 p. m., having rations, forage, and ammunition complete. There was no tentage, except the shelter-halves which some of the men had brought with them. Capt. Henry Marcotte, retired, the correspondent of the Army and Navy Journal, requested permission to accompany the detachment, which was granted, and soon all were *en route* for the front, entrusted with the task of opening the way for wheeled transportation and of demonstrating the practicability of the road for army wagons and field artillery.

For the first mile the road was excellent. It lay through one of the most fertile parts of the most fertile island in the world. A little stream trickling along the side of the road furnished plenty of water for both men and animals. At the end of the mile the detachment found a steep hill to descend. The Ordnance Department, which designed and built the carriage for the Gatling guns, had never foreseen the necessity for a brake, and it was therefore necessary to cut down bushes from the roadside and fasten the rear wheels by placing a stout pole between the spokes and over the trail of the piece. This locked the wheels, and the guns were thus enabled to slide down the steep hill without danger of a runaway. From this point the road became a narrow defile. The rank jungle closed in upon the trail, the long barbed leaves of the Spanish bayonet hung across and lacerated the legs of the mules until the blood trickled down to the hoofs; the boughs of the trees hung down over it so that even the men on foot had to stoop to pass under them, and the tortuous path winding in and out amid the dense tropical undergrowth made it impossible to see in places more than twenty-five or thirty yards ahead at a time.

The advance guard, consisting of all the members of one gun crew, had been organized at once upon starting, and this guard moved along the road about two hundred and fifty yards in advance of the detachment, scouting every path vigilantly to the right and left, and keeping a constant, careful lookout to the front. Their orders were, in case of encountering the enemy, to scatter in the underbrush, open fire with magazines, so as to produce the impression upon the enemy that there was a large force, and then slowly fall back upon the battery. The plan was, upon the first alarm, to bring the two leading guns into battery upon the road, with the fourth gun ready to be opened to either flank, while the gun crew of the third gun, which formed the advance guard, were to act as infantry support to the battery. It was hoped that the enemy would follow the advance guard as it retreated, and it was believed that the Gatling gun battery could take care of two or three regiments of Spaniards without help if necessary.

This form for the march had been adopted as the result of mature reflection. The general had offered a cavalry escort of two troops, and Gen. Sumner had rather urged the use of an escort, but it was desired to demonstrate that a battery of machine guns, properly manned and equipped, is capable of independent action, and does not need the assistance of either arm of the service. In fact, the Gatling gun men would have been rather pleased than not to have had a brush with the enemy without the assistance of either infantry or cavalry. But it was not to be.

The march was continued until darkness fell over the landscape, and the battery arrived at a beautiful camping-place about one mile east of Siboney, where a break in the water-pipe near the railroad track gave an ample supply of excellent water, and a ruined plantation, now overgrown with luxuriant sugar-cane, provided ample forage for the mules. The two troops of cavalry, which had been offered and refused as an escort, had reached this camping-place some time before, so that the wearied members of the detachment found pleasant camp-fires already throwing their weird lights and shadows over the drooping branches of the royal palm.

Here, in the midst of the jungle, they pitched their first camp in Cuba. The condition of the mules was duly looked to, their shoulders washed down with strong salty water, their feet carefully examined, and the animals then tethered to graze their fill on the succulent sugar-cane, after having had a bountiful supply of oats. Meantime the camp cooks had a kettle full of coffee simmering, and canned roast beef warming over the fire, and after a hearty meal the tired men stretched themselves upon the ground, with no canopy except the stars and only one sentinel over the camp, and slept more soundly than they had on board the tossing *Cherokee*.

CHAPTER 5

The March

At early dawn the battery arose, and, after a quick breakfast, resumed the march. Some half-mile farther on they passed a battery of light artillery which had preceded them on the road by some nine hours, and which had camped at this point awaiting forage. At Siboney the detachment stopped to look after the detail from the 13th Infantry, which had not yet reported. The detachment commander sought out the regimental adjutant, who referred him to the regimental commander, Col. Worth. This colonel was at first reluctant to allow the men to go, but, on being informed of the necessity for them, and after inquiring about the orders on the subject, he directed the detail to report immediately. All the members of this detail reported at once, except Corp. Rose, who had been left by his company commander on board ship.

The road from Siboney to the front was not known. There was no one in camp who even knew its general direction. Application was therefore made to Gen. Castillo, who was in command of a body of Cubans at Siboney, for a guide. After a great deal of gesticulation, much excited talk between the general and members of his staff, and numerous messengers had been dispatched hither and thither upon this important and very difficult business, a Cuban officer was sent with instructions to furnish a guide who could conduct the detachment to Gen. Wheeler's headquarters at the front. In the course of some twenty minutes, a dirty slouchy, swarthy, lousy-looking vagabond was pointed out as the desired guide, and was said to know every by-path and trail between Siboney and Santiago. He was told to go with

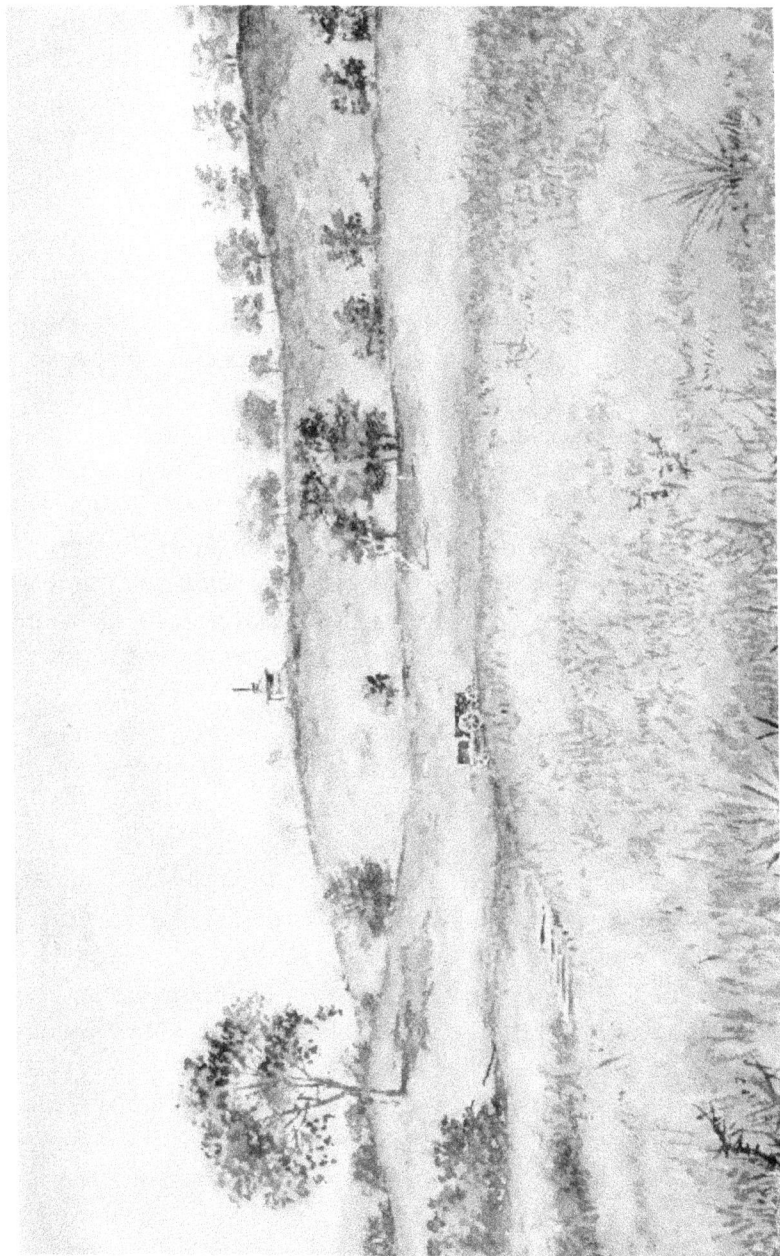

San Juan Hill

the detachment to Gen. Wheeler's headquarters and then return, and the detachment commander started for his command followed by his sable guide. Passing through a group of these brave Cuban heroes, he lost sight of his redoubtable guide for an instant, and has never since found that gentleman.

It would be just as well to add a description of the patriotic Cuban as he was found by the Gatling Gun Detachment during their campaign in behalf of Cuban independence, in the name of humanity; and this description, it is thought, tallies with the experience of all officers in the expedition.

The valiant Cuban! He strikes you first by his colour. It ranges from chocolate yellow through all the shades to deepest black with kinky hair; but you never by any chance see a white Cuban, except the fat, sleek, well-groomed, superbly mounted ones in "khaki," who loaf around headquarters with high-ranking shoulder-straps. These are all imported from the United States. They comprise the few wealthy ones of Spanish descent, who are renegade to their own nativity, and are appealing to the good people of the United States to establish them in their status of master of peons without any overlord who can exact his tithes for the privilege.

The next thing you notice is the furtive look of the thief. No one has ever yet had a chance to look one of these chocolate-coloured Cubans straight in the eye. They sneak along. Their gait has in it something of that of the Apache, the same soft moccasined tread, noiseless and always stealthy. Your impressions as to their honesty can be instantly confirmed. Leave anything loose, from a heavy winter overcoat, which no one could possibly use in Cuba, to—oh well, anything—and any Cuban in sight will take great pleasure in dispelling any false impressions that honesty is a native virtue.

Next you notice that he is dirty. His wife does sometimes make a faint attempt at personal cleanliness; this is evident, because in one bright instance a white dress was seen on a native woman, that had been washed sometime in her history. But as to his lordship, the proud male citizen of Cuba *libre*, you would utterly and bitterly insult him by the intimation that a man of his dignity ought ever to bathe, put on clean clothes, or even wash his hands. He is not merely dirty, he is filthy. He is infested with things that crawl and creep, often visibly, over his half-naked body, and he is so accustomed to it that he does not even scratch.

Next you observe the intense pride of this Cuban *libre*. It is manifested the very first time you suggest anything like manual labour—he is incapable of any other—even for such purposes as camp sanitation, carrying rations, or for any other purpose. His manly chest swells with pride and he exclaims in accents of wounded dignity, *"Yo soy soldado!"* Still his pride does not by any chance get him knowingly under fire. At El Poso some of him did get under fire from artillery, accidentally, and it took a strong provost guard to keep him there. If he ever got under fire again there was no officer on the firing-line who knew it.

He is a treacherous, lying cowardly, thieving, worthless, half-breed mongrel; born of a mongrel spawn of Europe, crossed upon the *fetiches* of darkest Africa and aboriginal America. He is no more capable of self-government than the Hottentots that roam the wilds of Africa or the Bushmen of Australia. He can not be trusted like the Indian, will not work like a negro, and will not fight like a Spaniard; but he will lie like a Castilian with polished suavity, and he will stab you in the dark or in the back with all the dexterity of a renegade graduate of Carlisle.

Providence has reserved a fairer future for this noble country than to be possessed by this horde of tatterdemalions. Under the impetus of American energy and capital, governed by a firm military hand with even justice, it will blossom as the rose; and, in the course of three or four generations, even the Cuban may be brought to appreciate the virtues of cleanliness, temperance, industry, and honesty.

Our good roads ended at Siboney, and from there on to Gen. Wheeler's headquarters was some of the worst road ever travelled. Part of it lay through deep valleys, where the sun was visible scarcely more than an hour at noontime, and the wet, fetid soil was tramped into a muck of malarial slime under foot of the mules and men. The jungle became ranker, the Spanish bayonets longer and their barbs sharper in these low bottom jangles. The larger undergrowth closed in more sharply on the trail, and its boughs overhung so much in some places that it became necessary to cut them away with axes in order to pass.

These guns were the first wheeled vehicles that had ever disturbed the solitude of this portion of Cuba. The chocolate-coloured natives of Cuba sneak; the white native of Cuba, when he travels at all, goes on horseback. He very seldom travels in Cuba at all, because he is not often there. Consequently the roads in Cuba, as a rule, are merely small paths sufficient for the native to walk along, and they carry the machete in order to open a path if necessary. These low places in the valleys were full of miasmatic odours, yellow fever, agues, and all the ills that usually pertain to the West Indian climate.

At other places the road ran along the tops of the foot-hills from one to two hundred feet higher than the bottom of these valleys. Here the country was much more open. The path was usually wide enough for the guns to move with comparative ease. Sometimes one wagon could pass another easily. These parts of the road were usually more or less strewn with boulders. The road was rarely level and frequently the upland parts were washed out. Sometimes it was only the boulder-clad bottom of a ravine; again the water would have washed out the gully on one side so deep as to threaten overturning the guns. The portions of the road between the valleys and the top of these foot-hills were the worst places the detachment had to pass. These ascents and descents were nearly always steep. While not at all difficult for the man upon horseback or for the man on foot, they were frequently almost too steep for draft, and they were always washed out. In places it was necessary to stop and fill up these washouts by shovelling earth and stone into the places before the detachment could pass.

On one of these occasions, while heaving rock to fill up a bad washout, Priv. Jones was stung by a scorpion. Jones did not know what had bitten him, and described it as a little black thing about as long as his finger. Fortunately there was a small supply of whisky with the detachment, and this remedy was applied to

Jones internally. Some soldier in the detachment suggested that a quid of tobacco externally would be beneficial, so this also was done. It was not a dressing favourable to an aseptic condition of the wound, perhaps, nor was there anything in the quid of tobacco calculated to withdraw the poison or neutralize its effects, so the doctors may characterize this as a very foolish proceeding; but country people skilled in simples and herb remedies might tell some of these ultra scientific surgeons that the application of a quid of tobacco or of a leaf of tobacco to the sting of a wasp or the bite of a spider, or even the sting of a scorpion, is nearly always attended by beneficial results. In fact, when Jones was stung there was a surgeon, a medical officer, who turned up even before Jones was treated with the whisky cure, and, upon receiving Jones' explanation that he had been heaving rock and had been bitten on the end of the finger by a little black thing, and after hearing the remarks of the men that it was very probably a scorpion sting, this medical officer very sagely diagnosed the accident to that effect, but was unable to prescribe any remedy because he had not brought along his emergency case. This medical officer, with his two attendant hospital satellites, had left both litter and emergency case upon the transport.

The ordinary line officer or soldier who is somewhat accustomed to carrying weights and does not require a hospital drill to teach him to carry a wounded comrade a few yards, looks with a certain degree of envy upon the possession of a hospital litter with its convenient straps for weight-carrying, and would consider this a very convenient means for carrying a pack. This litter is designed to enable two men, hospital attendants or band men, to pick up a wounded soldier weighing some 160 or 180 pounds and carry him from fifty yards to a mile if necessary, to a dressing-station or hospital shack. The medical field-case No. 1 weighs about sixty pounds filled, and field-case No. 2 weighs about forty pounds. These two cases contain all the medicines necessary to run a division hospital; the case of emergency instruments does not weigh above ten or twelve pounds, and would not be a burden for a child to carry. It is therefore difficult for the small-minded officer of the line to see why the

Medical Department was unable to have these medicines up at the front. They had the same means of locomotion provided for the other soldiers, by Nature, and they had, moreover, no particular necessity for all rushing to the extreme front. On the contrary, they had from the 23rd of June, when the landing began, at Baiquiri, until the 1st of July, to accomplish a distance of less than twenty miles; and it would seem reasonable that they might have had their medicine-cases up where they were needed by that time.

These gentlemen pose as the most learned, expert, scientific, highly trained body of medical men in the world. They are undoubtedly as well trained, as highly educated, and as thoroughly proficient as the medical officers of any army in the world. A summons of an ordinary practitioner would bring with him his saddle-bags of medicines; no physician in the city would pretend to answer even an ambulance call without having a few simple remedies—in other words, an emergency case; but it was an exception, and a very rare exception at that, to find a medical officer who took the trouble to carry anything upon his aristocratic back on that march to the front.

A conversation overheard between two medical officers on board a transport just before landing may serve to partially explain the state of affairs. Said surgeon No. 1 to surgeon No. 2, "We are going to land this morning; are you going to carry your field-case?" To which surgeon No. 2 indignantly replied, "No, I'm not a pack-mule!" Surgeon No. 1 again inquired, "Are you going to make your hospital men carry it?" To which surgeon No. 2 replied, "No; my men are not beasts of burden." Both of these medical officers went ashore; one of them had his field case carried; the other did not. Both of them were up at the firing-line, both did good service in rendering first aid. Both of them worked heroically, both seemed deeply touched by the suffering they were compelled to witness, and both contracted the climatic fever. But in the absence of medicines the role of the surgeon can be taken by the private soldier who has been instructed in first aid to the injured; for in the absence of medical cases and surgical instruments the first-aid packet is the only available source of relief, and these first-aid packets were carried by the private soldier, not by the Medical Department.

A little less "theory," a little less "science," a little less tendency to dwell on the "officer" part of the business, with a little more devotion to the duty of relieving suffering humanity—in short, a little less insistence upon "rank," would have vastly improved the medical service of the United States Army in the field at this time.

These remarks do not apply to the heroes like Ebert, Thorpe, Brewer, Kennedy, Warren, and a few others, who fearlessly exposed their lives upon the very firing-line. These men are the very "salt of the earth." The escape of even a "frazzle" of the 5th Corps was due to their superhuman energy and exertions. They did much to redeem the good name of their corps and to alleviate suffering.

But Priv. Jones recovered from the sting of the scorpion. In fact, soldiers were heard to exclaim that they would be glad to find a scorpion when they saw the character of the remedy applied in Jones' case.

The detachment left Siboney about 10 o'clock in the morning and tramped steadily along the road up hill and down until 12; then, upon finding a convenient place, it halted for dinner. The mules were unharnessed, coffee prepared, and, just as the detachment was about to begin this noonday meal, two of the peripatetic newspaper fraternity joined, *en route* to the rear. The ubiquitous correspondent had for the first time discovered the Gatling Gun Detachment, and they thought it was Artillery.

One of these gentlemen was a long, slim, frayed-out specimen of humanity, with a wearied and expressive droop of the shoulders; the other was a short, stout, florid, rotund individual, and his "too, too solid flesh" was in the very visible act of melting. The newspaper gentlemen were invited to participate in the noonday meal, and, with some gentle urging, consented. It was only after the meal was over that it was learned that this was the first square meal these men had had in over forty-eight hours. They had been with Gen. Wheeler at La Guasimas, had rejoined Wheeler after reporting that fight, in hopes of making another

WAGON TRAIN

"scoop," and were now on their way to Siboney, hoping to buy some provisions. Poor devils! They had worked for a "scoop" at La Guasimas; they had gone up on the firing-line and had sent back authentic accounts of that little skirmish; but they did not make the "scoop." The "scoop" was made by newspaper men who had remained on board the transports, and who took the excited account of a member of the command who had come back delirious with excitement, crazed with fear, trembling as though he had a congestive chill—who, in fact, had come back faster than he had gone to the front, and in his excited condition had told the story of an ambuscade; that Wheeler, Wood, and Roosevelt were all dead; that the enemy was as thick as the barbs on the Spanish bayonet; and that he, only he, had escaped to tell the tale. This was the account of the battle that got back to the newspapers in the form of a "scoop," and it was nothing more nor less than the excited imagination of the only coward who at that time or ever afterwards was a member of the famous Rough Riders. He was consequently returned to civil life prematurely.

The newspaper correspondent in Cuba was of a distinguished type. You recognized him immediately. He was utterly fearless; he delighted in getting up on the firing-line—that is, a few of him did. Among these few might be mentioned Marshall, and Davis, and Remington, and Marcotte, and King, and some half-dozen others; but there was another type of newspaper correspondent in Cuba, who hung around from two miles and a half to three miles in rear of the firing-line, and never by any possibility got closer to the enemy than that. The members of this guild of the newspaper fraternity were necessarily nearer the cable office than their more daring comrades; in fact, there were a few who were known to have been eight or nine miles nearer to the cable office during battles, and those correspondents were the ones who made the great "scoop" in the New York papers, by which a regiment that laid down and skulked in the woods, or ran wildly to the rear, was made to do all the fighting on the first day of July. This latter class of journalists were a menace to the army, a disgrace to their profession, and a blot upon humanity. Even the Cubans were ashamed of them.

The detachment resumed the march at half past 1, and encountered some very difficult road, difficult because it needed repairs. The most difficult places were the ascents and descents of the hills, and in nearly every case fifteen or twenty minutes' careful investigation was able to discover a means of getting around the worst places in the road. When it was not practicable to go around, J. Shiffer and his three fellow-teamsters would take a twist of their hands in the manes of their long-eared chargers, and apparently lift them down, or up, as the case might be, always landing on their feet and always safely. It was merely a question of good driving and will to go through. The worst places were repaired by the detachment before these reckless attempts at precipice-scaling were made. At one place there was a detachment of the 24th Infantry engaged in an alleged effort to repair the road. They did not seem to work with much vim. Chaplain Springer, having in the morning exhorted them to repentance and a better life and to doing good works unto their brethren, the enemy, was engaged at this point in the afternoon, it being Sunday, in a practical demonstration of what he considered good works. In other words, the chaplain, whose religious enthusiasm no one doubts, was engaged in heaving rocks with his own hands to show these coloured soldiers how they ought to make good road, and he was doing "good works."

It is but a just tribute to Chaplains Springer and Swift, of the Regulars, to say that they were conspicuous in the hour of danger at the point of greatest peril. In the fearless discharge of their holy office, they faced all the dangers of battle; nor did they neglect the care of the body while ministering to the spiritual needs of the soldiers. Springer, for example, collected wood and made coffee for all on the firing-line, within 400 yards of the block-house at El Caney; and Swift was equally conspicuous in relieving suffering, binding up wounds, and caring for the sick. There were probably others equally as daring; but the author knows of the deeds of these men, and desires to pay a tribute of respect to them. Chaplains of this stamp are always listened to with respectful attention when they express their views of the true course of life to obtain a blessed hereafter. They were in very sharp contrast to the long-visaged clerical gentlemen who were so much in evidence at Tampa, and who never got within 500 miles of danger.

The detachment safely passed all the bad places and obstacles in the road, arriving at Gen. Wheeler's headquarters about half past 4 o'clock, and reported. It was assigned a position between the advance outposts and directed to dispose of its guns in such a manner as to sweep the hills on which these outposts were placed. High hills to the right at a distance of about 2000 yards were supposed to be infested by the enemy, and a blockhouse which stood out against the sky-line was thought to contain a Spanish detachment. A high hill to the left at a distance of about 1000 yards had not yet been explored, and it was thought probable that some of the enemy was concealed on this hill also. The detachment commander was directed to report, after posting his battery, in which duty he was assisted by Col. Dorst, to Gen. Chaffee, who had charge of the outposts. The General inquired what the battery consisted of, and upon being informed that "It consists of four Gatling guns, posted so as to command the neighbouring hills," remarked in a very contemptuous manner, "You can't command anything." Gen. Chaffee subsequently had reason to revise his opinion, if not to regret the expression of it.

CHAPTER 6

The Battery in Camp Wheeler

At this point in the history of the detachment, it would be well to give some account of the reasoning which led to its formation and the personnel of the detachment.

Since the days of '65 the armies of the civilized world have adopted a rifle whose effective range is more than twice as great as that used in the Civil War. Very able discussions have been made upon the theoretical changes of the battle-field thus brought about, but no proper conclusion had been reached. It was acknowledged by all text-book writers that the artillery arm of the service would find much greater difficulty in operating at short ranges, and that assaults upon fortified positions would be much more difficult in the future. But only Gen. Williston, of the United States Artillery, had ever taken the advanced ground that in a machine gun arm would be found a valuable auxiliary as a result of these changed conditions. This theory of Gen. Williston's was published in the Journal of the Military Service Institute in the spring of '86, but never went, so far as Gen. Williston was concerned, beyond a mere theory; nor had the detachment commander ever heard of Gen. Williston's article until after the battle of Santiago.

A study of the science of tactics—not merely drill regulations, but tactics in the broader sense of manoeuvring bodies of troops upon the battle-field—had led Lieut. Parker to the conclusion that the artillery arm of the service had been moved back upon the battle-field to ranges not less than 1500 yards. This not because of lack of courage on the part of the Artillery, but as an inherent defect in any arm of the service which depends

GATLING BATTERY UNDER ARTILLERY FIRE AT EL POSO

upon draft to reach an effective position. It was not believed that animals could live at a shorter range in anything like open country. The problem of supporting an infantry charge by some sort of fire immediately became the great tactical problem of the battle-field. Admitting that the assault of a fortified position has become much more difficult than formerly, the necessity of artillery support, or its equivalent in some kind of fire, became correspondingly more important, while under the conditions it became doubly more difficult to bring up this support in the form of artillery fire.

The solution of this problem, then, was the principal difficulty of the modern battle-field; and yet, strange to say, the curtailed usefulness of artillery does not seem to have suggested itself to anybody else in the service previous to the first day of July. This problem had been made the subject of special study by him for several years, and had led to the conclusion that some form of machine gun must be adopted to take the place of artillery from 1500 yards down. This in turn led to the study of machine guns. The different forms in use in the different armies of the world had been considered, and it was found that there was none in any service properly mounted for the particular use desired. All of them required the service of animals as pack-mules, or for draft, while the very conditions of the problem required a gun to be so mounted that the use of animals could be dispensed with.

The Maxim gun has been reduced in weight to about 60 pounds, and is furnished with a tripod weighing about the same; but this is too heavy, and the supply of ammunition at once becomes a critical question. The Colt's automatic rapid-fire gun has been reduced to 40 pounds, with a tripod of equal weight, but here again the same difficulty presents itself. The soldier is capable of carrying only a limited amount of weight; and with his already too heavy pack, his three days' rations, together with the heat, fatigue and excitement of battle, it did not appear possible for any tripod-mounted gun to be effectively used.

The problem therefore resolved itself into the question of carriages: A carriage capable of carrying any form of machine gun using small-calibre ammunition, capable of being moved anywhere by draft, capable of being dismantled and carried on a pack-mule, and, above, all, capable of being moved by hand; required also some device for getting the requisite amount of ammunition up to the firing-line. A carriage and ammunition cart was invented fulfilling all these conditions and the invention was presented to the adjutant-general of the army for consideration, accompanied by a discussion of the proper tactical use of the gun so mounted. This discussion, in part, was as follows:

"It is claimed for this carriage that a machine gun mounted on it can be carried with a firing-line of infantry on the offensive, over almost any kind of ground, into the decisive zone of rifle fire and to the lodgement in the enemy's line, if one is made.

"On broken ground the piece can be moved forward by draft under cover of sheltering features of the terrain to a position so near the enemy that, under cover of its fire, an infantry line can effect a lodgement, after which the piece can be rushed forward by a sudden dash.

"The machine gun, mounted on this carriage, is especially adapted for service with the reserve of a battalion on the offensive, acting either alone or in regiment. Its use will enable the commander to reduce the reserve, thereby increasing the strength of the fighting-line, and yet his flanks will be better protected than formerly, while he will still have a more powerful reserve. If the fighting-line be driven back, the machine guns will establish a point of resistance on which the line can rally, and from which it can not be driven, unless the machine guns be annihilated by artillery fire.

"In case of counter-charge by the enemy, the superior weight and intensity of its fire will shake the enemy and so demoralize him that, in all probability, a return counter-charge will result in his complete discomfiture.

"Retiring troops as rear guards have in this weapon *par excellence* the weapon for a swift and sharp return with the power of rapidly withdrawing. If the enemy can by any means be enticed within its range, he will certainly suffer great losses. If he cannot be brought in range, his distance will be rather respectful."

This discussion as presented was entirely and absolutely original with the author and the result of his own unaided researches on the subject. It will be seen in the account of the battle how accurately the conditions there laid down were fulfilled.

But the carriage in use by the Gatling Gun Detachment was not the one proposed to the War Department. That carriage has not, as yet, been built, nor has the War Department in any way recognized the invention or even acknowledged the receipt of the communication and drawings.

The problem, therefore, confronting the Gatling Gun Detachment was to demonstrate the above uses of the machine gun, taking the obsolete artillery carriage drawn by mules, and endeavour to get the guns into action by draft. The personnel of the detachment alone accounts for their success. They got the guns up on the firing-line, not because of any superiority of the carriage over that in use by the artillery, for there was none; not because of aid rendered by other arms of the service, for they actually went into battle as far as 100 yards in advance of the infantry skirmishers; but because the Gatling Gun Detachment was there for the purpose of getting into the fight and was determined to give the guns a trial.

In the first place, all the members of the Gatling Gun Detachment were members of the Regular Army. All but three of them were natives of the United States, and those three were American citizens. Every man in the detachment had been selected by the detachment commander, or had voluntarily undertaken to perform this duty, realizing and believing that it was an extremely hazardous duty. Every member of the detachment possessed a common-school education, and some of them were well educated. All of them were men of exceptionally good character and sober habits. The drivers were Privs. Shiffer, Correll, Merryman, and Chase. The description formerly given of Shiffer applies, with slight modifications, to all the four. The first sergeant, Weigle, a native of Gettysburg, a soldier of eight years' experience in the Regular Army, a man of fine natural ability and good educational attainments, was worthy to command any company in the United States Army. Thoroughly well instructed in the mechanism of Gatling guns, of exceptionally cheerful and buoyant disposition, he was an ideal first sergeant for any organization. Steigerwald, acting chief of gun No. 1, was of German birth, well educated. He had chosen the military profession for the love of it; he was a man of wonderfully fine physique, a "dead sure" shot, and one who hardly understood the meaning of the word "fatigue." He was ambitious, he was an ardent believer in the Gatling gun, and he was determined to win a commission on the battle-field.

Corporal Doyle was a magnificent type of the old-time Regular—one of the kind that composed the army before Proctorism tried to convert it into a Sunday-school. In former days Doyle had been a drinking man; but the common opinion as expressed by his company officers even in those days was, "I would rather have Doyle, drunk, than any other non-commissioned officer, sober; because Doyle never gets too drunk to attend to duty."

Two years before this Doyle had quit drinking, and the only drawback to this most excellent non-commissioned officer had been removed. He was a thorough disciplinarian; one of the kind that takes no back talk; one who is prone to using the butt end of a musket as a persuader, if necessary; and Doyle was thoroughly devoted to the detachment commander. Corp. Smith was another of the same stamp. Corp. Smith loved poker. In fact, his *sobriquet* was "Poker Smith." He was one of the kind of poker-players who would "see" a $5 bet on a pair of deuces, raise it to $25, and generally rake in the "pot." It was Corp. Smith who thought in this Gatling gun deal he was holding a pair of deuces, because he didn't take much stock in Gatling guns, but he was a firm believer in his commanding officer and was prepared to "bluff" the Dons to the limit of the game.

Sergeants Ryder and Weischaar were splendid types of the American Regular non-commissioned officer, alert, respectful, attentive to duty, resolute, unflinching, determined, magnificent soldiers. Serg. Green was a young man, only twenty-three, the idolized son of his parents, in the army because he loved it; enthusiastic over his gun, and fully determined to "pot" every Spaniard in sight. Corp. Rose was like unto him. They were

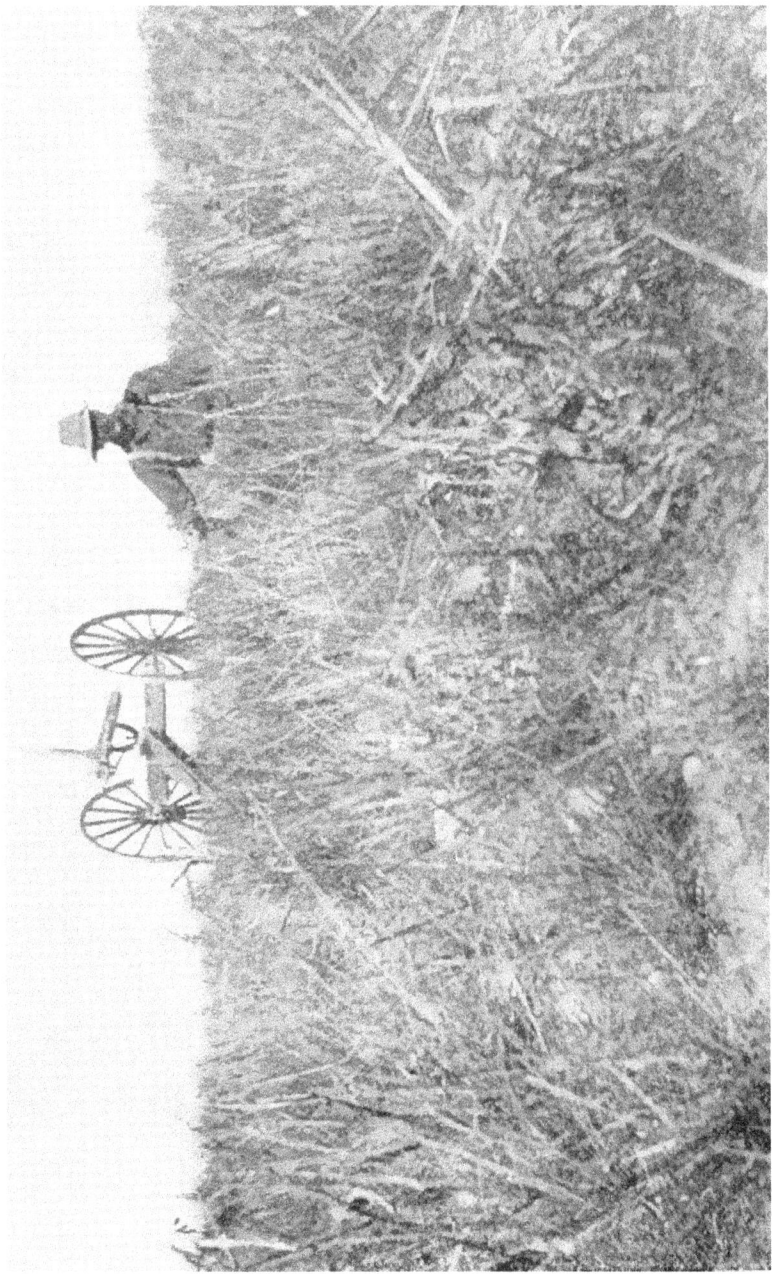

Gatling gun on firing line, July 1st

eager for nothing so much as a chance to get into action, and equally determined to stay there. The privates of the detachment were like unto the non-commissioned officers. They had volunteered for this duty from a love of adventure, a desire to win recognition, or from their personal attachment to the commanding officer; and there was not a man who was not willing to follow him into the "mouth of hell" if necessary. The gunners were expert shots with the rifle. Numbers 1 and 2, who turned the crank and fed the gun, respectively, were selected for their dexterity and coolness; the drivers, for their skill in handling mules; and each of the other members of the detachment was placed on that duty which he seemed best fitted to perform.

The roll of the detachment and its organization as it went into battle on the first day of July are subjoined:

GATLING GUN DETACHMENT
FIFTH ARMY CORPS

Commanding Officer, John H. Parker, first lieutenant, 13th Infantry. Acting First Sergeant, Alois Weischaar, sergeant, Co. A, 13th Infantry. Acting Quartermaster Sergeant, William Eyder, Co. G, 13th Infantry.

Gun No. 1: Acting Chief and Gunner, Charles C. Steigerwald, corporal, Co. A, 13th Infantry. No. 1, Private Voelker, Co. A, 12th Infantry. No. 2, Private Elkins, Co. D, 17th Infantry. No. 3, Private Schmandt, Co. G, 13th Infantry. No. 4, Private Needles, Co. E, 17th Infantry. No. 5, Private Click, Co. E, 17th Infantry. No. 6, Private Jones, Co. D, 13th Infantry. Driver, Private Shiffer, Co. E, 17th Infantry.

Gun No. 2: Chief, Sergeant William Ryder, Co. G, 13th Infantry. Gunner, Corporal Geo. N. Rose, Co. C, 13th Infantry. No. 1, Private Seaman, Co. B, 13th Infantry. No. 2, Private Kastner, Co. A, 13th Infantry. No. 3, Private Pyne, Co. H, 13th Infantry. No. 4, Private Schulze, Co. A, 17th Infantry. No. 5, Private Barts, Co. E, 13th Infantry. Driver, Private Correll, Co. C, 12th Infantry.

Gun No. 3: Chief, Sergeant Newton A. Green, Co. H, 13th Infantry. Gunner, Corporal Matthew Doyle, Co. C, 13th Infantry. No. 1, Private Anderson, Co. C, 12th Infantry. No. 2, Private Sine, Co. E, 17th Infantry. No. 3, Private Lauer, Co. C, 12th Infantry. No. 4, Private Dellett, Co. D, 17th Infantry. No. 5, Private Cory, Co. A, 13th Infantry. No. 6, Private Greenberg, Co. G, 13th Infantry. Driver, Private Merryman, Co. A, 17th Infantry.

Gun No. 4: Chief, Sergeant John N. Weigle, Co. L, 9th Infantry. Gunner, Corporal Robert S. Smith, Co. C, 13th Infantry. No. 1, Private McGoin, Co. D, 17th Infantry. No. 2, Private Misiak, Co. E, 13th Infantry. No. 3, Private Power, Co. A, 13th Infantry. No. 4, Private McDonald, Co. B, 17th Infantry. No. 5, Private Prazak, Co. E, 12th Infantry. Driver, Private Chase, Co. H, 13th Infantry. Cook, Private Hoft, Co. D, 13th Infantry. Assistant cook, Private Bremer, Co. G, 13th Infantry. Absent, sick, Private Murray, Co. F, 13th Infantry, at Tampa.

Sergeant Weigle was subsequently appointed first sergeant of Co. L., 9th Infantry, and of the Gatling Gun Detachment, vice Weischaar, relieved at his own request.

Another element which contributed much to the success of the detachment was the presence with it of Captain Marcotte. This excellent officer had served with great distinction in the Civil War, having been promoted from a private in the ranks through all of the grades up to a captaincy, for meritorious conduct in battle, and having failed of higher grades only because he was too badly shot to pieces to continue with the Army. He joined the detachment on the 25th of June, and his valuable advice was always at the disposal not merely of the commander, but of any member of the detachment who wished to consult him. He had spent seventeen years in the Cuban climate and was thoroughly familiar with all the conditions under which we were labouring. He contributed not a little, by his presence, his example, and his precept, to the final success of the organization. When the battery went under fire, Marcotte was with it. It was the first time most of the members

FORT ROOSEVELT

had passed through this ordeal, but who could run, or even feel nervous, with this grey-haired man skipping about from point to point and taking notes of the engagement as coolly as though he were sitting in the shade of a tree sipping lime-juice cocktails, a mile from danger.

Such was the personnel of the detachment. It lay in Camp Wheeler, which was only about a mile and a half from El Poso, where the first engagement occurred on the first of July, until that morning. The mules were daily harnessed up and drilled in manoeuvring the pieces, and the members of the detachment experimentally posted in different positions in order to get the most effective service.

On the 27th, Serg. Green was sent back to Siboney with orders to bring Corp. Rose or his body. He brought Corp. Rose, and the corporal was very glad to be brought.

The mules were fed with oats and on the juicy sugar-cane. It is worthy of mention that no other organization at the front had oats. A feed or two of oats was given to Gen. Wheeler and Col. Dorst for their horses; it was the first time their horses had tasted oats since leaving the transports, and was probably the last time until after the surrender. Furthermore, the Gatling Gun Detachment had "grub." Of course, it was "short" on potatoes, onions, and vegetables generally; these luxuries were not to be well known again until it returned to the United States; but it did have hardtack, bacon, canned roast beef, sugar, and coffee, having drawn all the rations it could carry before leaving Baiquiri, and was the only organization which had as much as twenty-four hours' rations. Gen. Hawkins and his whole brigade were living from hand to mouth, one meal at a time. The same was true of Gen. Wheeler and the whole cavalry division, and they were depending for that one meal upon the pack-mule train. On the 30th of June a complete set of muster- and pay-rolls, was prepared for the detachment, and it was duly mustered in the usual form and manner. It was the only organization at the front of which a formal muster was made, and was the only one there which had muster- and pay-rolls.

It rained on the 29th and 30th of June. Not such rains as the people of the United States are familiar with, but Cuban rains. It was like standing under a barrel full of water and having the bottom knocked out. These rains caused the rifles and carbines of the army to rust, and some quick-witted captain bethought himself to beg oil from the Gatling Gun Detachment. He got it. Another, and another, and still another begged for oil; then regiments began to beg for oil; and finally application was made for oil for a whole brigade. This led to the following correspondence:

Camp Six Miles from Santiago
29th June, 1898
The Adjutant-General
Cavalry Division, Present:
Sir,—I have the honour to inform you that I have learned that some of the rifles in this command are badly in need of oil, and that in some companies there is no oil to use on them. These facts I learned through requests to me for oil.

I therefore report to you that my men found at Altares (the second landing-place) and reported to me four (4) barrels of lard oil and three (3) barrels of cylinder oil, in an old oil-house near the machine shops.

If this be procured and issued, it will save the rifles and carbines from rust.

Very respectfully,
John H. Parker
Lt. Comdg. G. G. Detachment, 5th Corps

First Endorsement

Headquarters Cavalry Division
Camp 6 miles east of Santiago de Cuba
June 29, 1898
Respectfully referred to the adjutant-general, 6th Army Corps.
Jos. Wheeler
Major-General U. S. Vols., Comdg.

SERGEANT GREENE'S GUN AT FORT ROOSEVELT

Second Endorsement

Headquarters 5th Corps
June 29, 1898
Return. Lt. Parker will send a man back tomorrow to obtain the necessary oil.
By command of Gen. Shafter.
E. J. McClernand
A. A. G.

Third Endorsement

Headquarters Cavalry Division
June 29, 1898
Return Lt. Parker. Attention invited to the foregoing endorsement.
J. H. Dorst
Lieutenant-Colonel

Fourth Endorsement

June 30, 1898
The Quartermaster
Altares
Cuba
Please furnish to Sergeant Green of my detachment transportation for two (2) barrels of oil. He will show you an order from Gen. Shafter, and the matter is urgent. The soldiers must have this oil at once, as their rifles are rusting badly.
John H. Parker
Lt. Comdg. Gatling Gun Detach.

The quartermaster furnished the transportation and two barrels of oil were duly forwarded to the front and placed in charge of brigade quartermasters at different points, with orders to distribute out one quart to each company. This oil, perhaps, had some bearing upon the condition of the rifles in the fight following.

On the 27th of June, Captain Marcotte and the detachment commander made a reconnaissance of a high hill to the left of Camp Wheeler, and, having gained the top, reconnoitred the city of Santiago and its surrounding defences with a powerful glass, and as a result reported to Gen. Wheeler that the key of Santiago was the Morro *mesa*, a promontory or tableland overlooking the city on the east side at a distance of about a mile and a half and not at that time occupied by the enemy, with the proposition that a detail of a half-dozen men from the detachment should make a rush and capture this plateau, and hold it until the guns could be brought up. The general could not authorize the proposed undertaking, as it would have endangered the safety of his army, perhaps by leading to a premature engagement. By the time a sufficient reconnaissance had been made and convinced everybody of the value of this plan, the mesa had been strongly occupied by the enemy. It is still believed that the occupation of this height was practicable on the 27th of June, and thought, if it had been authorized, the Gatlings could have occupied and held this position against all the Spaniards in the city of Santiago.

CHAPTER 7

The Battle

On the 30th day of June, General Shafter pitched his camp
about half a mile in advance of Camp Wheeler in a valley, and
about five o'clock in the afternoon communicated the plan of
battle to the division commanders and to the commander of the
Gatling Gun Detachment.

Reconnaissance had developed the fact that the enemy occupied the village of El Caney, and that their first line of works
surrounded the city of Santiago at a distance of about a mile,
crowning a semicircular ridge. Between the position occupied
by the general's camp and this ridge, a distance of about two
and one-half miles, flowed the Aguadores and San Juan rivers,
and about one mile from the San Juan River, on the east side,
was a ruined plantation and mission house, called El Poso. Midway between El Caney and the Spanish position was a large
handsome mansion, called the Du Cuorot house, standing in
the midst of a large plantation and owned by a Frenchman,
which both sides had agreed to respect as neutral property. The
general plan of the battle as given to these officers on the 30th
of June was for one division of the army (Lawton's), assisted by
one battery of artillery (Capron's), to make an attack at day-
break upon the village of El Caney, and drive the enemy out
of it. Another division (Kent's) was to make an attack upon the
semicircular ridge of hills south of El Caney as soon as Lawton was well committed to the fight, both for the purpose of
preventing reinforcements from going to El Caney and to develop the enemy's strength. It was expected that Lawton would
capture El Caney about eight or nine o'clock in the morning,

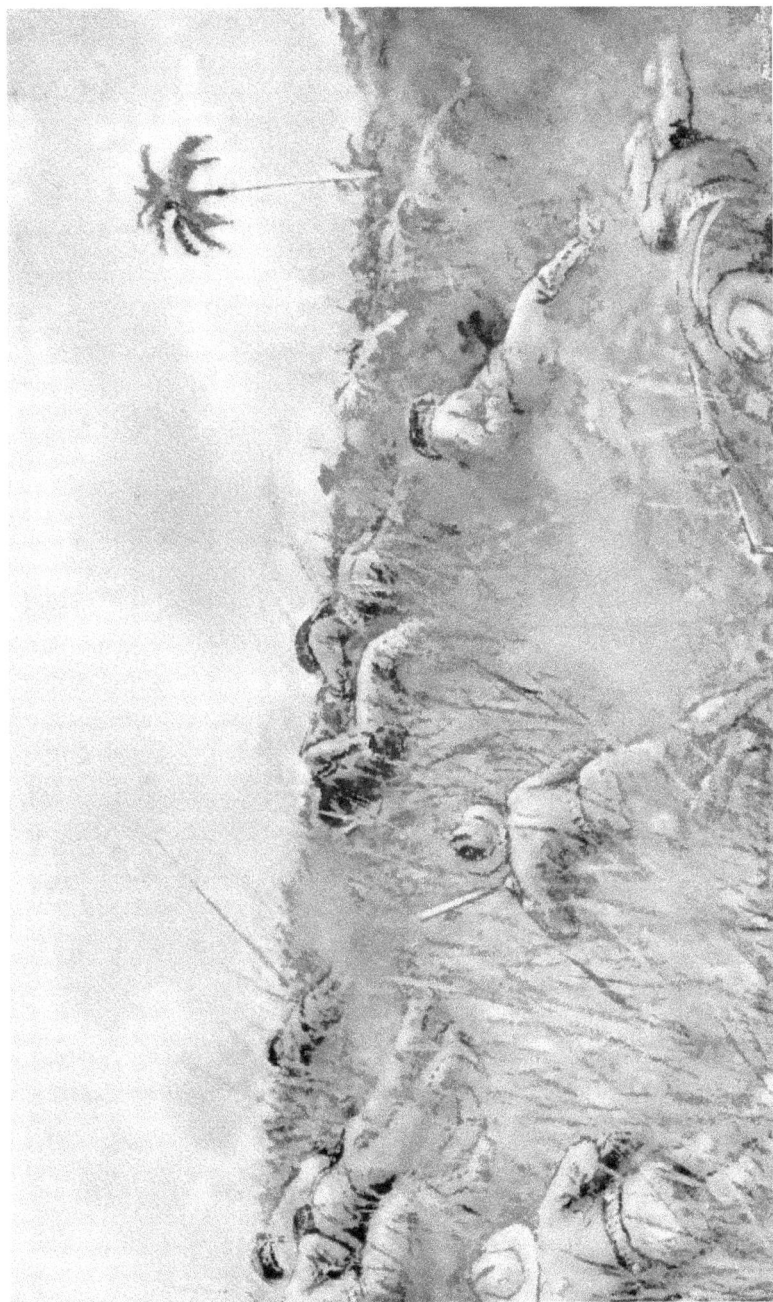

SKIRMISH LINE IN BATTLE

and pursue the retreating enemy, by the way of the Du Cuorot house, toward Santiago. This movement would cause Lawton to execute, roughly, a left wheel, and it was intended that in executing this manoeuvre Kent's right should join, or nearly join, Lawton's left, after which the whole line was to move forward according to the developments of the fight. Kent's attack was to be supported by Grimes' Battery from El Poso. The Gatling Gun Detachment was to move at daylight on the morning of July 1st, take position at El Poso sheltered by the hill, in support of Grimes' Battery, and there await orders.

This outline of the battle, as laid down by Gen. Shafter on the 30th day of June, was eventually carried out to the letter; its successful operation shut up a superior force in the city of Santiago, and compelled the surrender of the city.

Perhaps no better comment can be made upon the generalship of the corps commander, no higher compliment be paid, than the mere statement that he was able, fifteen hours before a shot was fired in the battle, to prescribe the movements of the different organizations of his command, and to outline the plan of battle as it was finally carried out, with a degree of precision which can be fully appreciated only by those to whom the plan was communicated in advance. In spite of slight changes, made necessary by local failures and unforeseen circumstances; in spite of the very poor cooperation of the artillery arm; in spite of the absence of cavalry, which made good reconnaissance practically impossible; in spite of the fact that he was operating against a superior force in strong entrenchments— the plan of battle thus laid down was finally carried out with perfect success in every detail.

The Gatling Gun Detachment was assembled at six o'clock, and so much of the plan of battle was explained to them as it was proper to give out, with orders that breakfast was to be prepared by four o'clock and the detachment be ready to move at 4:30. The plans were heard with careful attention by the men, and the wisdom of giving to them some idea of the work they were expected to do was fully vindicated on the following day, when they were compelled to lie nearly three hours under a dropping fire, waiting for "Lawton to become well engaged," after which the detachment moved forward, without a man missing, with the utmost steadiness and coolness, to the attack.

There was no nervousness displayed by the men. They knew their work was cut out for them, and each man was eager to play his part in the great drama of the morrow. There was no excited talk indulged in. None of the buzz of preparation nor the hum of anticipation which to the civilian mind should precede a desperate battle, but three or four members of the detachment took out their soldiers' hand-books and wrote in them their last will and testament, requesting their commander to witness the same and act as executor. The courage evinced by these men was not of that brutal order which ignores danger, but of the moral quality which, fully realizing that somebody must get hurt, quietly resolves to face whatever may happen in the performance of the full measure of duty.

At four o'clock the guard aroused the members of the detachment quietly, and each man found a good hearty breakfast waiting for him, consisting of hardtack, coffee, condensed milk, sugar, bacon, canned roast beef, and some canned fruit, which had been obtained somehow and was opened upon this occasion. It was the last square meal they were to have for several days. At half past four the camp equipage had all been packed upon the guns in such a manner as not to interfere with their instantly getting into action, and the battery started for the front.

The road to El Poso was very good and the mules trotted merrily along, preceded and followed by infantry also bound for the front. The Cubans, too, were in evidence; an irregular, struggling mob of undisciplined barbarians, vociferous, clamorous, noisy, turbulent, excited. Presently the Cubans and infantry in front of the battery halted and it passed beyond them, immediately throwing out the crew of the third gun in front as an advance guard. It reached El Poso at six o'clock, at which time there were no other soldiers there. The battery took position as directed, under cover in rear of the hill and to the right front of the El Poso house. The camp equipage and blanket-rolls, were removed and piled neatly upon the ground, and Priv. Hoft was

FORT ROOSEVELT

detailed to guard them, as well as one of the spare mules. About half past seven o'clock Grimes' Battery arrived, and Col. Mc-Clernand, the assistant adjutant-general of the corps. The battery of artillery halted upon the hill near the Gatlings, while its commander, the adjutant-general, the Army and Navy Journal correspondent, and the Gatling gun man climbed to the top of the hill to reconnoitre the enemy. They were accompanied by several attaches and a battalion of newspaper correspondents.

To the southwest, at a distance of about 3,000 yards, the city of Santiago lay slumbering in the morning sun. The chain of hills which surrounded the city, lying between it and our position, was crowned with rank tropical verdure, and gave no indications of military fortifications. There was no sign of life, a gentle land breeze swayed the tops of the royal palms, and the little birds flitted from bough to bough carolling their morning songs as though no such events were impending as the bombardment of a city and the death of 400 gallant soldiers. The gentle ripple of the creek, lapping over its pebbly bed at the foot of the hill, was distinctly audible.

The artillery officers produced their range-finders and made a scientific guess at the distance from the hill to a red brick building in the northern edge of Santiago. This guess was 2600 yards. They signalled to the lead piece of Grimes' Light Battery to ascend the hill. It was delayed for a moment while picks and shovels were plied upon the top of the hill to make slight emplacements for the guns, and at last, at ten minutes before eight o'clock, the first piece started the difficult ascent. The drivers stood up in their stirrups and lashed their horses and shouted; the horses plunged and reared and jumped. The piece stuck half way up the hill. The leaders were turned slightly to the right to give new direction and another attempt was made—ten yards gained. The leaders were swung to the left, men and officers standing near by added their shouts and blows from sticks. A tall artillery officer, whose red stripes were conspicuous, jumped up and down and swore; the team gave a few more jumps, then they wheeled the gun by a left about, with its muzzle pointing toward the city. It was quickly unlimbered and run to its place.

The second piece started up the hill. The drivers of this piece sat quietly in their saddles, and, with a cluck, started up the hill at a walk. The tall artillery officer shouted, and a driver muttered under his breath, "Damned fool!" Regardless of the orders to rush their horses, the drivers of this piece continued to walk up the hill. At the steepest part of the hill, they rose slightly in their stirrups, as one man, and applied the spur to the lead horses, and, at the same time, a lash of the quirt to the off horses of the team. The horses sprang forward, and in an instant the second piece was in battery. The third and fourth pieces were taken up in the same manner as the second.

The pieces were loaded; a party of newspaper correspondents produced their lead pencils and pads, and began to take notes; the little birds continued to sing. The Gatling Gun man, the Army and Navy Journal man, and the assistant adjutant-general stepped to the windward a few yards to be clear of the smoke. The range was given by the battery commander—2600 yards; the objective was named, a small, almost indistinguishable redoubt, below the hospital about 300 yards. The cannoneers braced themselves, No. 3 stretched the lanyard taut on his piece, and Grimes remarked, in a conversational tone, "Let her go."

The report of the field-piece burst with startling suddenness upon the quiet summer morning, and a dense cloud of greyish-coloured smoke spurted from the muzzle of the gun. Everybody involuntarily jumped, the sound was so startling, although expected. The piece recoiled eight or ten feet, and the gunners jumped to the wheels and ran it forward again into battery. Field-glasses were glued upon the vicinity of the brick hospital. There was a puff of white smoke and an exclamation, "A trifle too long!" The second piece was aimed and fired. There was no response. The third, and fourth, and fifth, with like results. It was like firing a salute on the Fourth of July. There was no indication of any danger whatever; laugh and jest were beginning to go round.

Suddenly a dull boom was heard from somewhere, the exact direction could not be located. The next thing was a shrill whistle overhead, and then a most startling report. The first Spanish shell exploded about twenty feet above the surface of the ground, and

A FIGHTING CUBAN AND WHERE HE FOUGHT

about twenty yards in rear of the crest of the hill. It exploded in the midst of our brave Cuban contingents, killed one and wounded several. The valiant sons of Cuba *libre* took to their heels, and most of the newspaper correspondents did likewise. The members of Grimes' Battery, who were not needed at the guns, were sent back to the caissons, and another round of shrapnel was sent in reply. Again a hurtling sound rent the air; again there was the fierce crack of a Spanish shell in our immediate vicinity, and, on looking around to see where this shell struck, it was observed that it had burst over the Gatling battery. Luckily, it had gone six or eight feet beyond the battery before exploding. A fragment of the shell had struck Priv. Bremer upon the hand, producing quite a severe contusion. The Missouri mules stamped the ground impatiently; one of them uttered the characteristic exclamation of his race, *"Aw! hee! aw! hee! aw!"* and the members of the detachment burst into a merry peal of hearty laughter. It was evident that this detachment was not going to run, and it was equally evident that the Missouri mules would stand fire.

A third shell whistled over the hill. This one burst fairly over Grimes' third piece, killed the cannoneer, and wounded several men.

The members of the detachment were now directed to lie down under their guns and limbers, except the drivers, who declined to do so, and still stood at the heads of their mules. Priv. Hoft, disdaining to take cover, shouldered his rifle and walked up and down, sentry fashion, over the pile of camp equipage.

Serg. Weigle, who had brought along a small portable camera, with a large supply of film-rolls, requested permission to photograph the next shot fired by Grimes' Battery. It was granted. He climbed to the top of the hill, stepped off to the left of the battery, and calmly focused his camera. Grimes fired another salute, and Weigle secured a good picture. A Spanish shell came whistling over the hill; Weigle, judging where it would burst from previous observations, focused his camera, and secured a picture of the burst. He then rejoined his detachment, and photographed it as it stood. He seemed chiefly worried for fear he would not get a picture of everything that happened.

GATLING CAMP AND BOMB-PROOFS AT FORT ROOSEVELT

The artillery duel continued for some twenty minutes. The infantry began to pass on, to the front. Grimes no longer needed the support of the Gatling guns, because he now had an infantry support in front of him, and was firing over their heads. Col. McClernand sent orders to the detachment to move to the rear, out of range. The order was obeyed.

Private Hoft, with the instinct of a true soldier, continued to tramp back and forth guarding the pile of camp equipage. The battery moved to the rear at a gentle trot, and, as it turned down the hill into the first ford by the El Poso house, a Spanish shell whistled over the head of Private Shiffer, who was leading the way, and burst just beyond his off mule. Shiffer didn't duck and nobody was hurt. Providence was taking care of this experiment. Corporal Doyle and two other members of the detachment got lost, and wandered off among the crowd of Cubans, but soon found the battery and rejoined. Orders were given that as soon as the battery was out of range, it should halt and face to the front, at the side of the road.

The battery halted about half a mile to the rear, and the 13th Infantry passed it here, on their way to the front. The comments bestowed were not calculated to soothe the ruffled feelings of people who had been ordered to retreat.

"I told you so."

"Why don't you go to the front?"

"Going to begin firing here?"

"Is this the place where you shoot?"

"Is this all there is of it?"

"I knew they would not get into the fight."

"Watch them hang around the rear."

"Going to start in raising bananas back here, John Henry?"

"What do you think of machine guns now?"

And similar remarks, of a witty but exasperating nature, greeted the detachment, from both officers and men, as the regiment passed on its way to the front. The only thing that could be done was to endure it, in the hope of getting a chance to make a retort later in the day.

About nine o'clock, the artillery firing ceased, and the Gatling

Gun Battery returned to El Poso. Grimes' guns were still up on the hill, but there were no cannoneers; they had ceased to fire, and had left their guns. Two or three dead men were lying on the side of the hill; wounded men were limping around with bandages. Cubans were again passing to the front. These fellows were trying to reach El Caney. They never got into the fight. They did reach the vicinity of El Caney, and the Spanish fired one volley at them. The Cubans set up a great howl, accompanied by vociferous gesticulations—and then "skedaddled."

During all this time the sound of firing had been heard toward El Caney. It had been opened up there about half an hour before Grimes first spoke at El Poso. The fire in this direction sounded like ranging fire, a shot every two or three minutes, and it was supposed that Capron was trying to locate the enemy. The sharp crack of musketry was heard on our front, it swelled and became continuous. It was evident that quite a fight was going on at El Caney, which was to our right about one mile and to our front perhaps half a mile. Kent's Division kept pushing forward on the El Poso road. Col. McClernand was asked for instructions for the Gatling Gun Detachment. He replied, "Find the 71st New York, and go in with them, if you can. If this is not practicable, find the best place you can, and make the best use of your guns that you can."

These were the only instructions received by the Gatling Gun Detachment until one o'clock.

The Gatling Gun Detachment moved forward about half a mile. They found the 71st New York lying down by the side of the road, partially blocking it. Troops passing them toward the front were compelled to break into columns of twos, because the road was crowded by the 71st. The colonel and his adjutant were sought and found, and informed of the detachment's instructions. Information was requested as to when and where the 71st was going into the fight. It appeared that they had a vague idea that they were going in on the left centre of the left wing. Lawton's Division at El Caney will be considered the right wing; Kent's Division and Wheeler's Division the left wing of the army at San Juan. The 71st did not seem to know

THE BETWEEN LINES SHOWING BULLET HOLES, THIS TREE GREW ON LOW GROUND

when it was going to move toward the front, nor just where it was going; and there was no apparent effort being made to get further down the road to the front. Wheeler's Division was also pressing forward on the road, dismounted cavalrymen, with no arms in their hands except their carbines without bayonets. With these same carbines these men were, a little later, to storm the entrenchments, manned by picked and veteran soldiers, who knew how to die at their posts.

With Wheeler's Division were the Rough Riders, the most unique aggregation of fighting men ever gathered together in any army. There were cowboys, bankers, brokers, merchants, city clubmen, and society dudes; commanded by a doctor, second in command a literary politician; but every man determined to get into the fight. About three-quarters of a mile in advance was the first ford, the ford of the Aguadores River; beyond this a quarter of a mile was another ford, the ford of the San Juan. The road forked about two hundred yards east of the Aguadores ford, turning sharply to the left. Down the road from El Poso crept the military balloon, it halted near this fork—"Balloon Fork." Two officers were in its basket, six or eight hundred feet above the surface of the ground, observing the movements of the troops and the disposition of the enemy.

The sharp crackle of the musketry began in front, and still the Gatling Gun Detachment lay beside the road with the 71st, waiting, swearing, broiling, stewing in their own perspiration, mad with thirst, and crazed with the fever of the battle. The colonel of the 71st was again approached, to ascertain whether he was now going to the front, but still there were no signs of any indication to move forward. So the long-eared steed was mounted and the ford of the Aguadores reconnoitred. The bullets were zipping through the rank tropical jungle. Two or three men were hit. Those who moved forward were going single file, crouching low, at a dog trot. There was no evidence of hesitation or fear here. Some of the "Brunettes" passed, their blue shirts unbuttoned, corded veins protruding as they slightly raised their heads to look forward, great drops of perspiration rolling down their sleek, shiny, black skins. There was a level spot, slightly open, beyond the ford of the Aguadores, which offered a place for going into battery; from this place the enemy's works on San Juan were visible, a faint streak along the crest of the hill illumined from time to time by the flash of Mausers.

On return to the battery, there were no signs of being able to enter the action with the gallant 71st, and, acting under the second clause of the instructions, the Gatling battery was moved forward at a gallop. Major Sharpe, a mounted member of Gen. Shafter's staff, helped to open a way through this regiment to enable the guns to pass. The reception of the battery by these valiant men was very different from that so recently given by the 13th Regulars.

"Give 'em hell, boys!"

"Let 'er go, Gallagher!"

"Goin' to let the woodpeckers go off?"

And cheer after cheer went up as the battery passed through. Vain efforts were made to check this vociferous clamour, which was plainly audible to the enemy, less than 1500 yards away. The bullets of the enemy began to drop lower. The cheering had furnished them the clew they needed. They had located our position, and the 71st atoned for this thoughtlessness by the loss of nearly eighty men, as it lay cowering in the underbrush near Balloon Fork.

Just before reaching the Aguadores ford, the battery was met by Col. Derby, who had been observing the disposition of the troops, from the balloon, and had afterward ridden to the front on horseback. The colonel was riding along, to push the infantry forward in position from the rear, as coolly as if on the parade-ground. A blade of grass had gotten twisted around a button of his uniform and hung down like a buttonhole bouquet over his breast. There was a genial smile on his handsome face as he inquired, "Where are you going?" and, on being informed of the orders of the detachment and of the intention to put the battery into action, he replied, "The infantry are not deployed enough to take advantage of your fire. I would advise that you wait a short time. I will send you word when the time comes." The advice was acted upon, the guns were turned out by the side of the road, and the men directed to lie down.

During the gallop to the front they had been compelled to run to keep up, there not being sufficient accommodation for them to all go mounted on the guns. They were panting heavily, and they obeyed the order and crept under the guns, taking advantage of such little shade as was offered. Troops continued to pass to the front. The crackle of musketry gradually extended to the right and to the left, showing that the deployment was being completed. More men were hit, but no complaints or groans were heard. A ball struck a limber-chest; a man lying on his face in the road, during a momentary pause of one of the companies, was perforated from head to foot: he never moved—just continued to lie there; the flies began to buzz around the spot and settle on the clotted blood, that poured out from the fractured skull, in the dust of the road. Down at the ford, some twenty-five or thirty yards in advance, men were being hit continually.

Shots came down from the trees around. The sharpshooters of the Spanish forces, who had been up in the trees during the artillery duel, and beyond whom our advance had swept, fully believing that they would be murdered if captured, expecting no quarter, were recklessly shooting at everything in sight. They made a special target of every man who wore any indication of rank. Some of our heaviest losses during the day, especially among commissioned officers, were caused by these sharpshooters. They shot indiscriminately at wounded, at hospital nurses, at medical officers wearing the red cross, and at fighting men going to the front.

The firing became too warm, and the Gatling battery was moved back about fifty yards, again halted, and faced to the front. It was now nearly one o'clock. The members of the detachment had picked up their haversacks on leaving El Poso, and now began to nibble pieces of hardtack. A bullet broke a piece of hardtack which a man was lifting to his mouth; without even stopping in the act of lifting it to his mouth, he ate the piece, with a jest.

Suddenly the clatter of hoofs was heard from the front. Lieut. Miley dashed up and said, "Gen. Shafter directs that you give one piece to me, and take the other three beyond the ford, where the dynamite gun is, find some position, and go into action." Sergeant Weigle's gun was placed at Miley's disposal, and the other pieces dashed forward at a dead run, led by the musical mule who uttered his characteristic exclamation as he dashed through the ford of the Aguadores.

The place formerly selected for going into action had been again twice reconnoitred during the wait, and a better place had been found about thirty yards beyond the ford of the San Juan River. The dynamite gun had stuck in the ford of the Aguadores; a shell had got jammed in it. The Gatlings were compelled to go around it. They dashed through the intervening space, across the San Juan ford, and up on the opening beyond. The position for the battery, partially hidden from the view of the enemy by a small clump of underbrush, was indicated. The right piece, Serg. Green's, was compelled to go into action in the middle of the road, and in plain sight of the enemy. While the pieces were being unlimbered, which was only the work of an instant, an inquiry was made of Captain Boughton, of the 3rd Cavalry, whose troop had just reached this point, as to the position of our troops and of the enemy, with the further remark that the battery had been under fire since eight o'clock, and had not seen a Spaniard. "I can show you plenty of Spaniards," replied Boughton, and, raising his hand, pointed toward the San Juan blockhouse and the ridge in its vicinity, sweeping his hand toward the right. It was enough. Before his hand had fallen to his side, the pieces were musically singing.

Corp. Steigerwald turned and asked, "What is the range, sir?"

To which was instantly replied, "Block-house, 600 yards; the ridge to the right, 800 yards," and Steigerwald's piece was grinding 500 shots a minute within a quarter of a second, playing upon the San Juan block-house.

Serg. Green took 800 yards, and began to send his compliments to the ridge beyond the block-house. In an instant Priv. Sine, at Green's gun, who was feeding, fell backward dead. At the same instant Priv. Kastner fell out. Sine was shot through the heart, Kastner through the head and neck. At this time Ryder's gun began to talk. It spoke very voluble and eloquent orations,

SPANISH BLOCK-HOUSE

which, although not delivered in the Spanish language, were well understood by our friends, the enemy, upon the hill.

Serg. Green, at the right gun, had run back for ammunition, and Corp. Doyle, when Sine fell, seized the pointing lever, and was coolly turning the crank while he sighted the gun at the same time. He was for the moment the only member of the detachment left at the piece, but was given assistance, and a moment later Green arrived and began to feed the gun.

Steigerwald was short-handed. Some of his men had been sun-struck during the run, and he, too, was compelled to work his gun with only one assistant. Then some of those who had been unable to keep up arrived at the battery and began to render assistance. Priv. Van Vaningham, who had gotten lost from his own command, began to pass ammunition. Priv. Merryman, who was holding his team back in the river, was impressed by a doctor to help carry wounded men, and Priv. Burkley, another man lost from his command, stepped into Merryman's place. Priv. Chase left his team, seeing the piece short-handed, and began to pass ammunition. The mules merely wagged their ears backward and forward and stamped on account of the flies.

All these changes were accomplished, and the pieces had not even ceased fire. Doyle had fed about 100 rounds, alone. Capt. Landis, of the 1st Cavalry, arrived just at this time, and volunteered to assist in observing the effect of the fire. He stood fearlessly out in the middle of the road, just to the right of Green's piece, in the very best position for observation, but, at the same time, a most conspicuous target for the enemy, and observed the effect of the Gatling fire, as though he were at target practice, reporting the same, continually, to the battery commander.

For the first two minutes the enemy seemed dazed, then suddenly a perfect hell of leaden hail swept through the foliage. The only thing that saved the battery from absolute destruction was that the enemy's shots were a little high. As it was, many of them struck the ground between the guns, and several hit the pieces. Three members of the detachment were slightly hurt. One mule was shot through the ear. He sang the usual song of the mule, shook his head, and was suddenly hit again on the fore leg. He plunged a little, but Priv. Shiffer patted him on the head and he became quiet. A bullet passed by Shiffer's head, so close that he felt the wind fan his whiskers, and buried itself in the saddle on the same mule. This sudden concentration of the enemy's fire lasted about two minutes.

About the same time the detachment heard a wild cheer start on the left and gradually sweep around to the left and right, until in every direction, sounding high above the din of battle and the crackling of the Mausers, even above the rattle of the Gatling guns, was heard the yell of recognition from our own troops. There was, for an instant, a furious fusillade on our right and left, and in a few moments the whole line of our troops had risen and were moving forward to the San Juan ridge. While moving forward, they necessarily almost ceased to fire, but the fire of the Gatlings continued, deadly and accurate. A troop of the 10th Cavalry, from our right and rear, came up, part of the squadron commanded by Col. Baldwin. Some of this troop did not understand the Gatling gun drama, and were in the act of firing a volley into our backs, when Lieut. Smith, who was to so heroically lose his life within ten minutes afterward, sprang out in front of the excited troopers, and, with tears in his eyes, implored them not to fire, that these were "our own Gatlings." They did not fire in our direction, but they did give a most thrilling and welcome cheer, as the squadron swept forward by our right. Col. Baldwin ran up, and shouted that he would place two troops in support of the battery as long as they were needed. It was the first time the battery had ever had a support of any kind.

After a couple of minutes, the enemy's fire perceptibly slackened. It was evident they were seeking cover from our fire in the bottom of their ditches, and our fire at this time was being made chiefly from the Gatling battery. This cessation of fire on the part of the enemy lasted about two minutes, and then the Gatling gunners observed the Spaniards climbing from their trenches. Until that time the Gatling battery had been worked with dogged persistency and grim silence, but from that mo-

SPANISH FORT OF THREE-INCH GUNS

ment every member of the battery yelled at the top of his voice until the command "Cease firing" was given. Groups of the enemy, as they climbed from their trenches, were caught by the fire of the guns, and were seen to melt away like a lump of salt in a glass of water. Bodies the size of a company would practically disappear an instant after a gun had been turned upon them.

This flight of the enemy from their trenches had been caused by the fact that the charging line had cut through the barb-wire fences at the foot of the hill, and had started up the slope. The Spaniards were unable to stay with their heads above the trenches to fire at the charging-line, because of the missiles of death poured in by the machine guns; and to remain there awaiting the charge was certain death. They did not have the nerve to wait for the cold steel. They were demoralized because they had been compelled to seek the bottom of their trenches. American troops would have awaited the charge, knowing that the machine gun fire must cease before contact could occur, but the Spaniards forgot this in their excitement, and made the fatal mistake of running.

The Gatlings had the range to perfection. Capt. Boughton, who was one of the first officers upon the hill, stated, on the 1st of September at Montauk, that he visited a portion of the Spanish trenches immediately upon arriving at the crest, and that the trenches which he inspected were literally filled with writhing, squirming, tangled masses of dead and wounded Spaniards, and that the edge of the trenches was covered with wounded and dead Spaniards, who had been shot in the act of climbing out. This execution was done mainly by the machine guns, because the infantry and cavalry were not firing much when it was done; they were running up the hill to the charge.

Colonel Egbert, who commanded the 6th Infantry, states, in his official report, that when his regiment reached the sharp incline near the top of the hill they were brought to a standstill because the Gatling bullets were striking along the crest. The officers of the 13th Infantry state the same thing. It was Lieut. Ferguson, of the 13th, who when the troops had climbed as high as possible under the leaden canopy which the Gatlings made to cover their charge, waved his white handkerchief as a signal to cease firing. At the same moment Landis exclaimed, "Better stop; our men are climbing the hill now." A shrill whistle gave the signal "Cease firing," and the Gatling Gun Battery, to a man, rose to their feet and gazed with absorbing interest as the long, thin, blue line swept forward and crowned the crest of the hill. An instant later an American flag floated proudly from the San Juan block-house; then the roar of musketry and the volley of rifles indicated that the fleeing enemy was receiving warm messengers as he ran down the hill toward his second line of entrenchments.

The next immediate duty confronting the detachment was to take stock of losses and to occupy the captured position in case of necessity.

Private Sine had been killed and Private Kastner was supposed to be mortally wounded. Private Elkins fell exhausted just as the Stars and Stripes were run up on the block-house. He had been knocked down by the pole of a limber, which struck him over the kidneys, but had continued to feed his gun until the very last. He was utterly exhausted. Sergeant Green had been wounded slightly in the foot, but not enough to disable him. Private Bremer had been hit early in the morning by the fragment of a shell on the hand. One or two other members had been merely touched, grazed by balls. Private Greenberg had been overcome by the heat. Merryman, one of the teamsters, as stated before, had been seized to carry wounded. Private Lauer was missing and Dellett sun struck. Private Hoft had joined the battery on hearing it go into action, and it was necessary to send someone back as guard over the camp equipage. A volunteer was called for, and it was with the utmost difficulty that a member of the detachment, Private Pyne, was induced to take this duty. He shot four Spanish sharpshooters, who were shooting at our wounded and our medical officers, out of trees near El Poso, during the remainder of the day. Private Chase had sprained his back so badly as to be unable to ride a mule; and two places were vacant for drivers. It was necessary to instantly supply this deficiency. Private Burkley, 16th Infantry, who had assisted in passing ammunition during the firing, volunteered to drive one of the teams, and Private Correll the other. Private Raymond, 6th Cavalry, and Private Van Vaningham, of the same regiment, also joined the detachment at this point, being separated from their own commands.

The pieces were limbered up as soon as these dispositions could be made, except Sergeant Ryder's gun, which had bent the pintle-pin and consequently could not be limbered quickly. The other two pieces and the limber belonging to Ryder's gun were moved forward on a run to the captured position on the San Juan ridge, gun crews riding or following as best they could. Both pieces went into action on the right of the road. A limber was then sent back for Ryder's gun, and it was brought up, Priv. Shiffer performing this duty under a perfect hail of dropping fire. In advancing from the position at the ford to the captured position it was necessary to cut three barb-wire fences. The members of the detachment behaved with the utmost coolness, all working together to remove these obstructions, and not a man sought shelter, although a dropping fire was striking around the detachment, from some source. Where this fire came from it was impossible to tell; but it did not come from the enemy.

The two pieces which first reached the top of the hill were halted under shelter of the crest, while the ground above was reconnoitred. It was instantly observed that the enemy was coming back for a counter-charge. Accordingly the pieces were immediately run to the top of the hill, the drivers, Shiffer and Correll, riding boldly up and executing a left-about on the skirmish line, where the skirmishers were lying down. The pieces were unlimbered and instantly put into action at point-blank range, the skirmishers giving way to the right and left to make way for the guns. The enemy was less than 300 yards away, and apparently bent on recovering the position.

The fire immediately became very hot. A skirmisher, who had thought to gain a little cover by lying down beside the wheel of the right gun (Green's), was shot through the arm. "I knowed it," he growled; "I might have knowed that if I got near that durned gun, I'd get potted." He rolled down behind the crest; a soldier produced an emergency packet, staunched the blood, and the wounded soldier, finding no bones broken, returned to the firing-line and resumed his work. The enemy, at this part of the line, began to waver and again broke toward his second line of entrenchments.

Just at this moment, Lieut. Traub came up and shouted, "Gen. Wood orders you to send one or two of your guns over to help Roosevelt." The order to move the guns was disregarded, but Traub pointed out the enemy, which was menacing Col. Roosevelt's position, and insisted. About 600 yards to the right, oblique from the position of the guns and perhaps 200 yards, or less, in front of the salient occupied by Col. Roosevelt and the 3rd Cavalry (afterward called Fort Roosevelt), there was a group of about 400 of the enemy, apparently endeavouring to charge the position. There was no time to notify the second piece. Serg. Green's gun was instantly turned upon this group, at point-blank elevation. The group melted away. Capt. Marcotte states that, after the surrender, some Spanish officers, whom he met, and who were members of this group, described this to him, stating that the enemy seen at this point was a body of about 600 escaping from El Caney; that they were struck at this point by machine gun fire so effectively that only forty of them ever got back to Santiago; the rest were killed.

Serg. Green's gun, already heated to a red heat by the continuous firing of the day, had been worked to its extreme limit of rapidity while firing at this body of the enemy, and on ceasing to fire, several cartridges exploded in the gun before they could be withdrawn. A ball lodged in one barrel from one of these explosions, and this piece was drawn down out of action just as the piece which had been left at the ford returned. Subsequently the disabled piece was sent back to the ford, with the idea that that would be a safer place to overhaul it than immediately in rear of the firing-line. The piece remained at the ford until the night of the 3rd of July, when it was brought up to the battery, then at Fort Roosevelt, and on the 4th was finally overhauled and put into action. This led to the impression, on the part of some of the command, that one of the Gatlings had been blown up,

TENTACE IN CUBA

which was not true. The gun was not injured, except that one barrel could not be used during the remainder of the fighting, but the gun was used on the morning of the 4th, and during the whole of the engagement on the 10th and 11th, as well as on outpost duty, using nine barrels instead of ten.

Following this repulse of the enemy, which occurred about 4:30 p. m., there was a lull in the firing. Advantage was taken of this to visit Col. Roosevelt's position and inspect the line of battle. Upon reaching the salient, Col. Roosevelt was seen walking up and down behind his line, encouraging his men, while a group of them was held, just in the rear of the crest, in charge of Maj. Jenkins, to support the firing-line if necessary. On the right of the Rough Riders, the 3rd Cavalry were in the fight, and Capt. Boughton was again encountered.

The firing suddenly began again, and it was remarkable to observe the coolness with which these two officers sauntered up and down the line, utterly regardless of the bullets, which were cutting the grass in every direction. There were no soft places on this part of the hill. The enemy's sharpshooters, up in high trees, were able to see every point of the crest, and were dropping their shots accurately behind it at all points.

Just at this moment, Serg. Weigle came up with his gun. Serg. Weigle had had a hard time. His gun had been taken, under direction of Lieut. Miley, to a point near the San Juan farmhouse, and pulled to the top of the hill. Weigle, whose only idea of a battle, at this time, was a chance to shoot, had been, to his intense disgust, restrained from opening fire. Then the piece had been taken down from the hill and around to the left of the line, where Lieut. Miley's duty as aide had carried him, to observe the progress of the battle, and Weigle had been again denied the privilege of "potting" a Spaniard. He was the most disgusted man in the American Army; he was furious; he was white-hot; he was so mad that the tears rolled down his cheeks, as he reported with a soldierly salute, "Sir, Serg. Weigle reports, with his gun. Lieut. Miley did not allow me to open fire. I would like to have orders."

In spite of the critical condition of the engagement, it was extremely ludicrous; but the reopening of the fire at this moment presented an opportunity to accommodate the sergeant to his heart's content. He was directed to run his piece up on the firing-line, report to the officer in charge thereof, and go into action as soon as he pleased. Within thirty seconds he was getting his coveted opportunity. He fired until his gun became accidentally jammed, pulled it down behind the crest of the hill and removed the defective cartridge, returned it and repeated this operation, actually bringing the gun down three times, and returning it into action, doing very effective work, and all the time displaying the utmost coolness and good judgment. A sharpshooter began to make a target of Weigle's gun, and "potted" a couple of men belonging to the cavalry near it. This made Weigle so mad that he turned the gun, for a moment, upon the tree in which the sharp-shooter was concealed. That sharpshooter never shot again. Finally, Weigle's gun got so hot, and he himself so cool, that he concluded the piece was too warm for further firing. So he ran it down behind the hill, and ran his detachment back on the hill with rifles, and, during the remainder of the evening, the members of this crew practiced with "long Toms" upon the Spanish soldiers.

On returning to the other two pieces near the road, they were moved to another position, on the other side of the road. This precaution was judicious in order to conceal the pieces, or change their position, because the enemy had pretty thoroughly located them in the previous brush, and it was rather dangerous to remain at that place. It was now nearly sundown. Scarcely had the pieces opened at this new position, when a battery of the enemy's artillery, located near the hospital, began to fire at them. There was a heavy gun, which made a deep rumbling sound, and this sound was supplemented by the sharp crack of a field-piece. A shell came whistling overhead and exploded within thirty yards of the battery, just beyond it. Another one came, and this time the enemy's artillery was located. Quick as a flash, the two Gatlings were turned upon the enemy's guns at the 2000-yard range. Another shell came whistling along and exploded about ten feet overhead and twenty feet in rear of

After the rain

the battery. It tore up the grass in rear of the battery. After this engagement was over, Priv. Shiffer picked up the still hot fuse of this last shell. It was a large brass combination fuse, and set at eight seconds, which justified the estimated range. This third shell was the last one the enemy was able to fire from these pieces. The powerful field-glasses which were used in locating the battery revealed the fact that as soon as the Gatling guns were turned on it, the Spanish gunners ran away from their pieces. The big gun turned out to be a 16-centimeter converted bronze piece, mounted on a pintle in barbette, rifled and using smokeless powder. It was also found that they were firing four 3-inch field-pieces of a similar character in this battery, as well as two mountain guns.

It is claimed that this is the first time in the history of land fighting that a battery of heavy guns was ever put out of action by machine-gun fire. This battery of the Spanish was never afterward able to get into action. Their pieces, which had been loaded for the fourth shot, were found on the 18th of July, still loaded, and a Spanish officer gave the information that they had lost more than forty men trying to work that battery, since the 1st of July. This is accounted for by the fact that this Spanish battery was made the subject of critical observation by the Gatling Gun Detachment from this time on.

During this last engagement it had been necessary to obtain more men to assist in carrying ammunition, and Capt. Ayers, of the 10th Cavalry, had furnished a detail, consisting of Serg. Graham and Privates Smith and Taylor, Troop E, 10th Cavalry. These coloured soldiers proved to be excellent. They remained with the battery until the end of the fighting on the 17th, and were in every respect the peers of any soldier in the detachment. Serg. Graham was recommended for a medal of honour. Privates Smith and Taylor did as good service, were as willing, as obedient, as prompt, and as energetic in the discharge of their duties as any commanding officer could wish to have. It is a great pleasure to be able to give this testimony to the merits of our coloured troopers, and to say, in addition, that no soldiers ever fought better than the "Brunettes" of the 9th and

10th Cavalry, who fought from the 3rd of July until the 12th, near or with the Gatlings.

After the firing at the ford had ceased, Capt. Marcotte had returned to El Poso to investigate the movements of our artillery. These were then, and have remained, one of those inscrutable and mysterious phenomena of a battle; incomprehensible to the ordinary layman, and capable of being understood only by "scientific" soldiers. The charge upon the San Juan ridge was practically unsupported by artillery. No American shells had struck the San Juan block-house; none had struck or burst in its vicinity; not even a moral effect by our artillery had assisted in the assault. So Marcotte had gone to investigate the artillery arm.

He returned at sundown, and brought the information that our baggage was safe at El Poso; that Private Pyne, still alive and unhurt, had been doing good work against the enemy's sharpshooters; and, better than all this, had brought back with him a canteen of water from the San Juan River and a pocket full of hardtack. He poured out his hardtack, and it was equally distributed among the members of the detachment, each man's share amounting to two pieces. Each man was also given a sup of water from the canteen, and this constituted their only supper on that night, as they had been compelled to throw away everything to keep up with the guns.

Having disposed of that, exhausted Nature could do no more; they lay down in the mud where they stood, and slept so soundly that even the firing which occurred that night did not arouse them from their slumbers. They were not disturbed until Best's Battery began to occupy this hill about four o'clock in the morning. They were then aroused and the Gatling guns were drawn down, and the whole battery moved to the salient occupied by the Rough Riders, because their position was at that time closest to the enemy, and, as was determined by the previous day's reconnaissance, offered a chance to enfilade several of the enemy's trenches with machine gun fire.

To dispose of the subject of artillery, it may be said that Best's Battery and some other artillery occupied the ground vacated by the Gatlings on the morning of July 2nd, fired four shots, and then withdrew with more haste than dignity. They remarked, "This is the hottest fire to which artillery has been subjected in modern times," and lit out to find a cooler place. They found it—so far in rear that their fire was almost equally dangerous to friends and foes on account of the close proximity of the two firing-lines. The obvious conclusion is that machine guns can live at close ranges, where artillery can not stay. There is no better light artillery in the world than that which had to withdraw from San Juan block-house and its vicinity, on the morning of July 2nd.

CHAPTER 8

Analysis of the Battles at Santiago

The situation of affairs on the night of the 1st of July was rather critical. The plan which the general had laid down had been delayed in execution at El Caney, while the impetuosity of the troops had precipitated an unexpected rapidity of movement at San Juan. Capron's Battery had opened at El Caney about half past seven o'clock, with badly aimed and ill-directed fire, which did very little damage to the enemy. The troops engaged in this part of the battle were pushed forward until, by about eleven o'clock, they had become pretty thoroughly deployed around the vicinity of Las Guamas Creek. They had also extended slightly to the right and to the left toward the Du Cuorot house. The Spanish forts obstinately held out, and the handful of Spanish soldiers in El Caney and vicinity stubbornly resisted the attack made by our troops.

About nine o'clock, Hamilton's right piece, No. 3 of Capron's Battery, succeeded in planting a shell directly in the old stone fort, which knocked a hole in the masonry; but, just at this juncture, the battery was ordered to cease firing at the blockhouse, and to shell the enemy's trenches. The enemy forthwith utilized the hole made in the wall by the shell as a loop-hole, and continued to fire through it until the fort was taken by the infantry assault at about half-past four o'clock. No worse commentary than this could possibly be made upon the tactical handling of this battery of artillery, because, having obtained perfectly the range of the enemy's stronghold, it was simply asinine not to knock that block-house to pieces immediately.

So Lawton's Division had remained in front of El Caney,

held by about 1000 Spaniards, while the shadows crept from the west to the north, from the north to the northeast, and from the northeast toward the east. It was coming toward night before the artillery was finally turned loose. One corner and the roof of this block-house were knocked off, but even then the artillery was so poorly handled that the enemy had to be dislodged from this block-house by hand-to-hand fighting, A single Hotchkiss gun, properly handled, should have converted it into ruins in thirty minutes.

While these events were transpiring, Kent and Wheeler, constituting the left wing of the army, had moved forward on the El Poso road, parallel to the Aguadores River, as far as the San Juan had captured the San Juan farm-house, and had gradually deployed to the right and to the left along the San Juan River. About one o'clock their line had swept forward and had captured the first ridge between the San Juan and the city of Santiago, the "San Juan ridge," driving the enemy on this portion of the field into their last trenches. But the right flank of this wing was entirely unsupported, and the road by the way of Fort Canosa to San Juan, passing by the portion of the line subsequently occupied by the dynamite gun, marked the extreme position of the right of this wing of the army. The enemy was already well toward its right, and had the excellent El Caney road to move upon. He was thoroughly familiar with the country, while the troops composing this wing were exhausted by the charge. This wing had no reserve that the firing-line knew of, and, as a matter of fact, had none except two battalions of the 71st New York, which had not got into battle, and which were scattered along the road from the San Juan River to Siboney.

The position occupied by the left wing of the army was a strong natural position, but had no protection for the right flank. In this, Lawton's Division did not execute the part of the battle assigned to it. Thus the officers on the San Juan ridge, who knew anything about the plan of the battle, were constantly directing their gaze, at every lull in the fighting, toward El Caney, and to the right of Gen. Wood's position, but there were no indications of the approach of Gen. Lawton.

Returning now to the right wing: the San Juan block-house and the ridge in its vicinity having been captured. This capture occurred at 1:23½ p. m. The Spanish commander at El Caney had been killed about noon, his men had suffered heavily, and the new commanding officer discovered that his retreat by the El Caney road was threatened. The only other line of retreat was by way of the San Miguel and Cuabitas roads. The Spanish forces at El Caney were also running low in their ammunition, and it was therefore decided to withdraw. Portions of the Spanish troops did withdraw, some by way of the El Caney road toward Santiago; the remainder, some 350 or 400, were crushed in the final charge upon El Caney, between 4 and 4:30 o'clock.

Gen. Lawton's Division then proceeded down the El Caney road to Santa Cruz, passing by way of the masonry bridge. This was about dusk. The division marched in columns of fours, with the artillery in front in column of sections, and without even an advance guard thrown out. The artillery had passed the masonry bridge and had nearly reached the Santa Cruz farm-house, when the order was given to halt. The division halted in the road and began to cook supper. Fires were kindled, and coffee put on to boil. Suddenly, a few shots came scattering over the ridge and dropped in among the troops. A messenger was sent back to Gen. Shafter to inform him that further advance in this direction was not practicable, as the enemy had been encountered in force. The position this division was destined, in the beginning, to occupy was within less than 300 yards of where it halted. There was no large body of Spanish troops in that portion of the field. The whole valley between that ridge and Santiago had been swept by machine gun fire during the afternoon. It is possible that there might hare been a few Spanish pickets on the ridge, but this is not believed to be probable. There was some firing about this time from the Spanish trenches near Fort Canosa, at the 13th Infantry upon the hill where the dynamite gun was subsequently placed. This was the firing that alarmed Lawton's Division and caused the report mentioned to be sent back to General Shafter.

This statement of the conditions has been necessary in order

NATIVE INDUSTRY

to understand why the counter-march was made by Lawton's Division. The position at El Caney had ceased to be of any importance as soon as the San Juan block-house and ridge were taken; any Spanish troops remaining at El Caney were necessarily victims. But it was vitally important to hold the position gained by the left wing. The appearance of a heavy force of the enemy in front of the masonry bridge could signify only one thing, and that was that the left wing, with its right flank in the air, was liable to be doubled up at any moment by a heavy force of the enemy striking it upon that flank. Further, that Gen. Lawton, with this column advancing on the El Caney road as before explained, was liable to be struck at the head of his column and similarly doubled up. The enemy would thus interpose between the two wings of the army, cutting Lawton off, and probably defeating the army in detail, unless something be done immediately.

Of course, it is known now that this operation of the enemy was never probable for an instant; but that was the status of affairs at midnight on July 1st, as then reported to the commanding general.

Lawton was, therefore, ordered to withdraw, by way of the El Caney road, back to Gen. Shafter's headquarters in rear of El Poso, from which position his division was rushed forward on the El Poso road to San Juan on the 2nd of July. His men were marched almost all night, almost all day the next day, and were well-nigh utterly exhausted when they reached a position in rear of the right flank of the left wing. It was supposed, up to this time, at headquarters, that the information on which this marching was ordered was correct.

During the time that Lawton had been countermarching from Santa Cruz, back by way of El Poso, there had been, as before stated, no reserve for the left wing. The independent division of Gen. Bates had been ordered to the front as rapidly as possible. Part of it had reached the vicinity of El Poso, and from there one or two of the regiments had participated in the fight, late on July 1st; but nobody on the firing line knew anything about Bates' independent division at this time, and it was too much exhausted to be useful as a reserve. The morning of the 2nd it was used to extend the lines. It is therefore evident, now that the history of the battle is understood, that the Gatling guns were the only effective reserve which the left wing of the army had during the night of July 1st and all day on the 2nd.

Acting on this belief, the Gatling Gun Battery was placed in reserve, in the rear of Fort Roosevelt, on the morning of July 2nd, and was held there in reserve all day on July the 2nd and 3rd. The pieces were placed within twenty yards of the firing-line, just below the crest of the hill. The feed-guides were filled, and the gun crews lay down beside their pieces. The battery was ready to either support the firing-line against a charge, or protect its flank against a turning movement. But it was not considered necessary or desirable to run the pieces up on the firing-line in the open, and participate in the trench-firing, which was the only fighting done on July 2nd and 3rd. It was considered that the battery was too valuable as a reserve to sacrifice any of its men uselessly. Some very well-meaning officers urged that the battery be rushed up on the hill and put into action, but this was stubbornly refused, under the third clause of the instructions given on the 1st of July, "to make the best use of the guns possible." Gen. Wood and Col. Roosevelt were consulted, and they concurred with the above views, and the battery remained in reserve.

On the morning of the 2nd of July a handsome young soldier, in the uniform of a Rough Rider, approached the battery commander, saluted, and said, "Col. Roosevelt directs me to report to you with my two guns." Inquiry elicited the fact that the young trooper was Serg. William Tiffany, that he had command of two Colt's automatic rapid-fire guns, with a crew consisting of Corp. Stevens and six men, and that he had 4,000 rounds of 7-millimeter ammunition. Four thousand was not a very large supply for two guns which could fire at the rate of 500 shots each per minute. Fortunately, the Gatling Gun Detachment had found time, on the 1st of July, to collect about 10,000 rounds of Mauser ammunition in the captured trenches, and a comparison of the Mauser with the 7-millimeter ammunition at once

Charge on San Juan hill

disclosed the fact that it was precisely the same ammunition which Tiffany had brought along for his guns. The problem of ammunition supply for Tiffany's guns was solved. He now had 14,000 rounds, and his guns became a very powerful reinforcement at this point.

Serg. Tiffany and his men had carried these guns from Siboney to the firing-line upon their backs. How they got the four boxes of ammunition through they themselves could hardly tell. The firing was too heavy to mount the tripods in the trenches during the daytime, so placing the guns was deferred until night. For some reason it was not practicable to place the tripods on the night of the 2nd, and they were finally placed on the night of the 3rd; Serg. Tiffany, with two of his men, aiding in digging the emplacements.

While digging, suddenly a burst of firing broke out, and it was believed by many that a serious night attack had been made. During the firing, Capt. Ayers, of the 10th Cavalry, and Col. Roosevelt again displayed those characteristics of fearless bravery which so endeared these two gallant officers to their men. Some of the troops in the trenches had begun to fire wildly. In fact, all the firing was done wild; there was no sense in any of it; there was no occasion for it. Intent listening to the enemy's fire made it absolutely certain that their firing never approached nearer our lines. There may have been some small body seeking to explore the road, but there was no indication of any attack in force. At any rate, Roosevelt and Ayers determined to stop the firing of our line, and suddenly, above the din of battle, these two officers could be heard, tramping up and down the trench in front of their men, haranguing, commanding, ridiculing their men for shooting in the dark. Ayers told his men that they were no better than the Cubans, upon which the burly black troopers burst into a loud guffaw, and then stopped firing altogether. Roosevelt told his men that he was ashamed of them. He was ashamed to see them firing valuable ammunition into the darkness of the night, aiming at nothing; that he thought cowboys were men who shot only when they could see the "whites of the other fellow's eyes." They also stopped firing. The enemy's bullets continued to whiz by for a few minutes, and they too ceased firing, and everybody began to laugh at everybody else. Tiffany had joined the two officers in their walk up and down, exposing himself with the utmost coolness. He and his men now succeeded in placing his guns in the trench, and, from that time until the end of the fight, they could hardly be induced to leave them long enough to eat; they didn't leave them to sleep— they slept in the trench by the guns.

About one o'clock on the 3rd there was a lull in the firing, during which a flag of truce was sent with a communication to General Toral, notifying him that a bombardment would follow unless he surrendered. The firing was resumed and continued until about half past twelve on the 4th of July, at which time another flag of truce went up, and there was no more firing until the 10th of July at about three o'clock. Troops, however, were compelled to lie on their arms; the relief was constantly in the trenches, and the nervous strain was even worse than the actual dangers of battle.

Negotiations for capitulation having failed, firing was resumed about three o'clock on the 10th, and continued until one o'clock on the 11th of July. In this firing all four of the Gatling guns were used; Tiffany's guns and the dynamite gun under Serg. Borrowe participated. Three of the Gatling guns had been placed in the trench on the night of July 3rd. The wheels were taken off and laid on the ground in the rear of the pieces; sand-bag revetments were built up in front of the guns, and each crew divided into two reliefs. One relief was required to be constantly at the gun and always ready for instant action. The fourth gun, the one that had been temporarily disabled, was repaired on the 4th, thoroughly cleaned, and placed in reserve behind the crest of the hill. On the 4th of July, Serg. Borrowe had been directed to obey any instructions given him by the Gatling gun commander, and the dynamite gun had been placed in position to cooperate with the battery of machine guns. There were now, therefore, seven pieces in the battery. It was the most powerful and unique battery ever used in battle.

The Sims-Dudley pneumatic dynamite gun throws a Whitehead torpedo, carrying a charge of four and one-half pounds of explosive gelatine; the effective force of this charge is equal to that of nine pounds of dynamite, No. 1. The charge explodes, on striking, by means of a percussion fuse, and steadiness of flight is secured by means of a vane. The propelling force is a charge of seven ounces of smokeless powder. The gun is pointed in the same manner as a mortar, and fired in the same manner as a field-piece. During the 10th and 11th considerable attention was devoted to the tactical cooperation of the guns composing this unique battery.

The plan adopted was for the dynamite gun to throw a shell toward a designated point. Upon the explosion of this shell the Spanish soldiers invariably exposed themselves, and were immediately assailed by machine gun fire. Occasionally a dynamite shell would fall with sufficient accuracy to do efficient work on its own account. On the afternoon of the 10th a dynamite shell fell in a long trench near Fort Canosa, clearing out the trench. The Spanish survivors were cut down almost to a man by the machine gun fire, and the Spanish troops were unable to occupy this trench until the following morning, when the operation was repeated, practically destroying the usefulness of this trench during the whole fight. Capt. Duncan, of the 21st Infantry, states that this relieved his battalion of an enfilading fire, and was a valuable service to them. Another dynamite shell, on the afternoon of the 10th, fell into a Spanish battery of artillery, near the brick hospital, and completely destroyed the battery, which consisted of two 3-inch guns. In all, about a dozen dynamite shells were thrown with some degree of accuracy, and with good effect.

The fourth Gatling gun, which had been held in reserve, was used during the afternoon of July 10th, and all day on the 11th, to pour a vertical fire upon the city of Santiago, beyond that portion that was visible to the American troops. Perhaps 6,000 or 7,000 shots were thus dropped into the heart of the city, making the streets unsafe, communication difficult, and striking terror to the hearts of the Spanish troops who were held there in reserve. Gen. Toral, in his official statement to his own government, specifically mentions this fire as one of his principal reasons for surrender.

GATLINGS AT BAIQUIRI JUST BEFORE STARTING FOR THE FRONT

On the afternoon of the 10th and during the 11th of July a battery of mortars, under command of Capt. Ennis, posted about half a mile to the right of the machine gun battery, threw a few shells at the enemy's entrenchments. There were four of these mortars in action and they were placed behind the ridge in a perfectly safe position. They threw, perhaps, twenty-five shells all told. The first eight or ten failed to explode for the reason that the fuses had not been punched. Finally, Capt. Ennis discovered that his shells were not exploding, and, on inquiry, found that there was no fuse-punch in the battery. He succeeded in finding a brad-awl, which, luckily, some member of the battery had in his pocket, and showed a sergeant how to punch the fuse with a brad-awl. After this the mortar shells exploded all right. None of this fire, however, was directed at the city; it was directed at the trenches of the enemy, and not over eight or ten of the shells fell with any precision. The mortar fire was effective in the sense that it tended to demoralize the enemy, but its material effect was very small.

There was no firing of field-pieces during all this time of which any account is necessary. The field-pieces were even less useful during this time than they were on the 1st of July, if such a thing could be possible.

On the night of the 4th of July the reserve Gatling gun was posted to command the Fort Canosa road, in support of a picket on that road, and from that time until the surrender this piece was posted there every night. The members, therefore, of this detachment did practically double duty. This was the gun in charge of Sergts. Weischaar and Ryder, referred to in the official report. Luckily, it was not fired, but there can be no doubt of the immense value it would have had if its use had been necessary.

Summing up the use of machine guns from the 2nd to the 11th of July, inclusive, it may be said that they demonstrated the use of the arm as a tactical reserve and an auxiliary to an outpost, and that, in combination with a dynamite gun, they demon-strated that a new arm of the service had been formed which can live at closer range to the enemy, and do far more effective work, than artillery. Nor is this all to be considered. It should be

135

remembered that a field-piece throws a shell which breaks into 273 fragments. The machine gun throws 1000 shots, and each of these shots is aimed with absolute precision. Therefore, at any effective range, the machine gun is far superior to a field-piece against anything except material obstacles. Of course the machine guns will not do to batter down stone walls, nor to destroy block-houses. It had already been demonstrated on the 1st of July that "machine guns can go forward with the charging-line to the lodgement in the enemy's position," and that "their presence on the field of battle, with a supply of ammunition for ten minutes, is a decisive factor in the engagement."

These were the principal points claimed for the machine gun in the discussion of the subject on the 1st of January. The use of the machine gun for advance and rear guards was not demonstrated at Santiago, for the reason that no opportunity was presented.

CHAPTER 9

The Volunteers

The white flag went up at one o'clock on the 11th, and this was the end of the fighting at Santiago. The Rough Riders had been moved from the hill at Fort Roosevelt to a position west of the El Caney road, and one of the Gatling guns had been sent with them. This gun was brought back on the 17th after the surrender. Various other movements of troops occurred before the 17th, which had been decided upon by the generals as the last day of grace.

Gen. Toral had been notified that one o'clock on the 17th was the time for either the surrender or the signal for the assault. The hour approached, and still the Spaniard attempted to delay.

The orders for the assault were issued.

The troops lay in the trenches with their fingers on the triggers. Gen. Randolph had come and pushed the artillery into better positions. The pieces were loaded and the gunners stood with their lanyards in their hands. The ammunition-boxes were opened. The nervous tension of the line was terrific. The troops on the extreme right and left, designated for the assault, were only waiting the word to dash forward upon the entrenchments of the enemy.

Then suddenly from Gen. Wheeler's headquarters a mounted officer was seen spurring eastward along the crest. He was waving his hat over his head. His horse gathered speed, and the foam began to fly from his flanks and nostrils, and as Capt. McKittrick passed he called, "No cheering, please; the city and province of Santiago have surrendered."

The members of the Gatling Gun Detachment walked to

the top of the hill, and, facing toward the gallant enemy who had so valiantly defended the foredoomed city, silently took off their hats.

All along the line the reception of the glorious news was marked by comments upon the gallant defence which had been made. There was no demonstration which could have hurt the feelings of so magnificent a foe. Five minutes after the surrender the American trench was lined by American troops on our side and Spanish troops on the other. The Spanish troops brought bottles of *mescal*, *aguardiente*, and wine. Our troops carried hardtack and canned roast beef. These recent foes began at once to exchange the necessaries of life and souvenirs of the siege of Santiago. They fraternized as all brave men do after the battle. A few Cubans skulked around the rear of our line, despised by both sides.

The next day witnessed the formal surrender of the city. At twelve o'clock, the preliminary formalities having been complied with, the 9th Infantry and a battalion of the 13th Infantry, the two regiments which had been adjudged first honours in the assault, were ready as an escort to raise the flag in the heart of the city. All of the other regiments were formed upon the ground which they occupied during the siege. As the second-hands of our watches showed the minute of twelve, noon, a field-piece burst upon the stillness of the sultry day, and the band began the strains of The *Star-spangled Banner*. Every hat was taken off, and an instant later, efforts to restrain it being ineffectual, six miles of solid cheering encompassed the latest American city.

Grizzled old soldiers, scarred with wounds from Indian wars, and gay recruits who had arrived too late to join in the fighting, grey-haired generals and athletic young subalterns, all forgot propriety and the silence usually enjoined in ranks and joined in that tremendous yell. From over on the right of the El Caney road we could hear the *"Rah! rah! rah!"* of Harvard and the *"Rah! rah! rah!"* of Yale, mingled with the cowboy yell of the Indian Territory. From the ranks of the Regulars came the old Southern yell, mingled with the Northern cheer. The most thrilling and dramatic moment of the Spanish-American War had passed into history.

CUBAN CART USED BY GATLING GUN DETACHMENT, PRIVATE J. SHEFFER DRIVING

The troops settled themselves down to wait for developments, and while they waited, opportunities were presented for the first time to make observations of the personnel of this heterogeneous army.

The American Regular is a type of his own, and no description of him is necessary. He was the fighting strength of the 5th Corps. Only three Volunteer regiments participated in the charges of July 1st–the 71st New York, the 2nd Massachusetts, and the 1st Volunteer Cavalry.

The Volunteers presented many different types: some good, some otherwise. There should be no sympathy with that servile truckling to popular sentiment which speaks of our brave Volunteers indiscriminately, as if they were all good and all equally well instructed. There were Volunteers who were the equals of the Regulars in fighting and in leadership. And there were some who should have been at home pulling on a nursing-bottle or attending a kindergarten. To praise them indiscriminately creates a false impression on the public, and works a rank injustice toward those who were really good and efficient in the service. It does even worse than that: it fosters the popular idea that all there is to do to make soldiers is to take so many labourers, clerks, hod-carriers, or farmers, and put on them uniforms, arm them with rifles, and call them "gallant Volunteers"! Out upon such an insane delusion!

Fighting is a scientific trade. It would be no more absurd to give an idiot a tambourine and call him a musician—he would be an idiot all the same. So with the clerk, the labourer, the hod-carrier, the teacher; he remains the same in spite of all the polished arms, resplendent uniforms, and pompous titles bestowed upon him. He remains just what he was before, until he learns his new trade and becomes a soldier by the acquisition of the necessary knowledge and experience to practice his new calling.

It is one of the duties of trained officers to tell these homely truths to the people who have not made a study of the matter, in order that they who foot the bills may understand what they pay for and why they do it. And it is equally the duty of the citizen who has no knowledge of the subject to give a fair hearing to such statements, and, if he finds them correct after due investigation, to translate the information thus imparted into such laws as will in future supply an army composed of soldiers who can fight, instead of a herd of ignorant incompetents who die like rotten sheep within half an hour's ride by rail of their own homes.

These remarks can be illustrated by observations in Cuba.

For example, the 34th Michigan pitched its camp on the hill at Fort Roosevelt on the 2nd of August. They were in an awful condition. A man had died in one company the day before, and there had not been enough able-bodied men in the company to bury him. A detail had to be made from another company to dig the grave. More than fifty per cent of the regiment were sick, and the remainder were far from well. At this time, more than two weeks after the surrender, they were still cooking individually. Within fifteen minutes after their arrival they were overrunning the Gatling gun camp, picking up the firewood which had been gathered by the detachment for cooking purposes. An attempt to stop this marauding was received with jeers. A green-looking Wolverine at once began to make catcalls, and was ably seconded by his comrades. Sentinels were then posted over the Gatling gun camp, with orders to keep the Michiganders out; they abused the sentinels in the same manner, and their officers made no effort to restrain them. It became necessary to make a personal matter of it, which was promptly done, and one Wolverine was thereafter respectful—so respectful, in fact, that he jumped to attention and took off his hat to even the privates of the detachment.

The regiment took a delicate revenge. They had dug neither latrines nor sinks. Up to this time they used the surface of the camp-ground over their own camp for this purpose. They now took possession of a trench within twenty yards of the battery's tents. The nuisance was intolerable, and was reported to their brigade headquarters. No attention was paid to the report. Twelve hours later it was again reported, with the same result. Twelve hours after this it was a third time reported, with the same result. In the meantime not a single shovelful of dirt had been thrown on the trench and an odour arose from it which was not exactly like the perfume of "Araby the blest."

Forty-five hours after the arrival of the regiment notice was served upon the brigade commander thereof that, unless the nuisance was abated immediately, a sentinel would be placed over the offending ditch and notice would be given to General Bates, the division commander, requesting the action of an inspector; notice was further served that if any resistance were made, four Gatling guns would be turned loose upon the 34th Michigan and the regiment swept off the face of the hill and into Santiago Bay for a much-needed bath. It was enough.

Officers and men ran instantly for spades and proceeded to fill up the trench. Report was then made to Gen. Bates, the division commander, of the offence and action had thereon, with the information that the Gatling gun commander awaited to answer any complaints. An investigation was immediately made, with the result that such action was sustained.

There were some ignorant Volunteers at Santiago, but of all the wilful violation of all the laws of sanitation, camp hygiene, and health ever seen, these particular Volunteers did the most outrageous things. They threw their kitchen refuse out on the ground anywhere; half of the time they did not visit the sink at all, but used the surface of the ground anywhere instead; and they continued these offences at Montauk Point. They raked over an abandoned camp of the Spanish prisoners on their arrival at Fort Roosevelt, and appropriated all the cast-off articles they could find, using the debris for bedding. This surgeon, a "family doctor" from the pine woods in northern Michigan, did not seem to regard these matters as of any importance. His attention was called to them, but he took no action. In short, there was no law of health which these people did not utterly ignore, no excess dangerous to health which they did not commit. Three-fourths of them were too sick for duty, and the rest looked like living skeletons. They fairly wallowed in their own filth—and cursed the climate of Cuba on account of their sickness.

In sharp contrast to the 34th Michigan was the 1st U. S. Volunteer Cavalry, the Rough Riders. This was an organization the

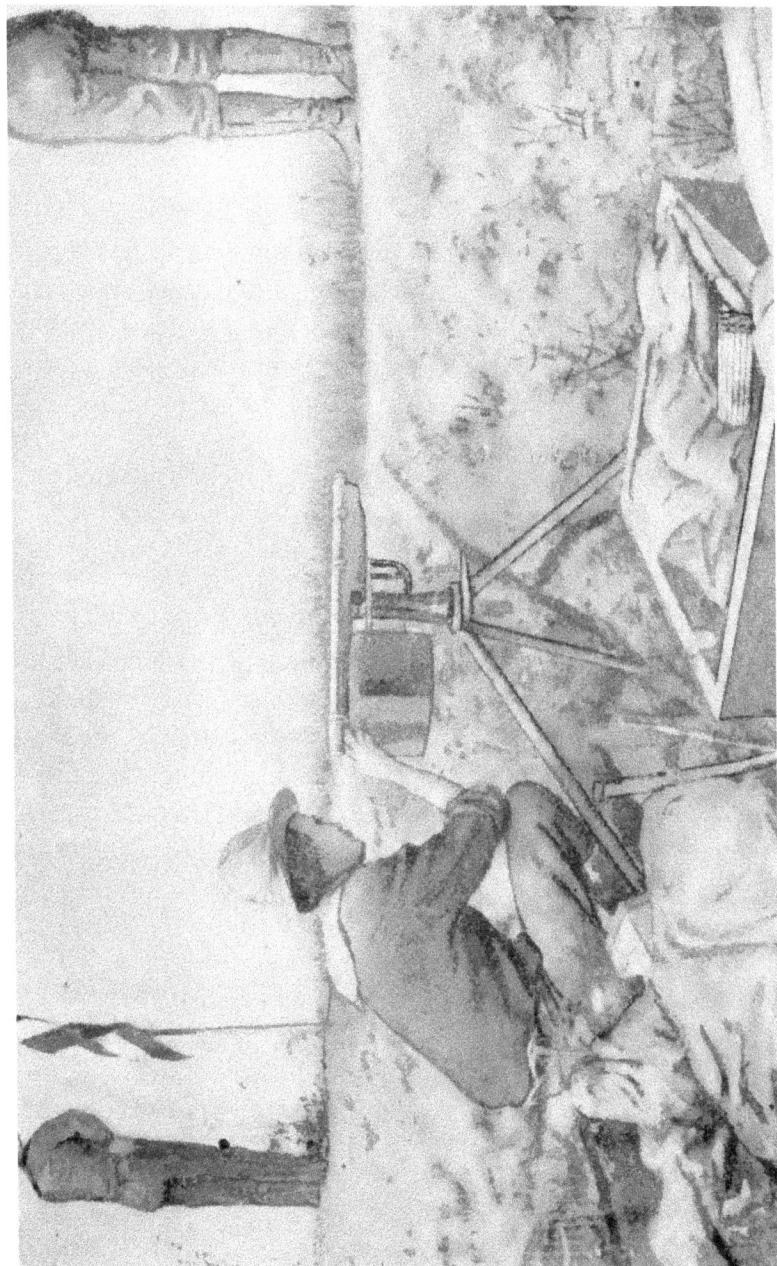

TIFFANY AT HIS GUN IN THE TRENCH

peer of any in the Regular Army in morale, in fighting, and in every quality that goes to make up a fine body of soldiers. They were picked men; all classes were shown in that organization. The tennis champion was a private, the champion oarsman of Harvard a corporal. On the 2nd of July a stockbroker of Wall Street who can sign his check for $3,000,000 was seen haggling with a cow-puncher from the Indian Territory over a piece of hardtack. Both were privates and both were fine soldiers. The whole regiment was just such a medley, but fought like Regulars, and endured like Spartans. They hung on like bull-dogs, and charged like demons. They were as strict about the camp police as Regular Army surgeons, and as punctilious about saluting as a K. O. on "official relations."

Withal, they were a clean-mouthed, clean-clad, clean-camped lot of gentlemen, each in his way, from the "Hello, pard!" of the cowboy to the frozen stare of the monocled dude from Broadway. And they fought—like Regulars; there is no other just comparison.

Roosevelt said: "They are the 11th Cavalry."

He found enthusiastic endorsers of this remark in every Regular who saw them fight. They were the finest body of Volunteers who ever wore uniform, and they were stamped indelibly with the personality of Theodore Roosevelt. Pushing, aggressive, resolute, tenacious, but self-contained, cool, and restrained, they represented the very best type of what the Volunteer ought to be—but often was not.

Above them all, however, shone out three types.

Theodore Roosevelt. He needs no eulogy from my pen. He has done everything, and in each occupation has been conspicuously successful. He is, however, a born soldier. His virile frame contains the vigorous mind, the keen intellect, the cool judgment, and the unswerving, never-hesitating courage of the natural soldier. He is affable and courteous, or stern and scathing, as circumstances demand. One instant genial smiles overspread his expressive countenance, whereon the faintest emotion writes its legend with instantaneous and responsive touch; the next, on occasion, a Jove-like sternness settles on his face, and, with a facility of expression bewildering to less gifted tongues, scathing invective, cutting sarcasm, or bitter irony impress upon an offender the gravity of a breach of discipline. Withal, he is modest. He appreciates his own power, but there is no undue display of that appreciation, no vainglorious boasting over achievements which read like a fairy-tale. Fittest to lead or follow, idol of every true soldier. Who, that knows him as those who fought beside him, does not wish to see him at the head of that army and that nation of which he is the brightest ornament in every position, civil, military, or political?

Woodbury Kane—social leader, Fortune's favourite, aristocratic, refined, cultured, wealthy, *haut ton de haut ton*, and *sabreur sans peur et sans reproche*—how shall I paint him to you as I learned to know him in those dreadful, delightful seventeen days in which we lived only from instant to instant, and every man unconsciously bared his soul to his comrades because he could not help it?

A gentleman—he always looked that in the fullest sense of the word. Well groomed; in those days when our bed was a mud-puddle and our canopy the stars, when the music which lulled us to sleep was the hum of the Mauser bullets and the vicious popping of the Remingtons, when water to drink had to be brought at the peril of life for every mouthful, Kane turned up every morning clean-shaved and neatly groomed, shoes duly polished, neat khaki, fitting like a glove and brushed to perfection, nails polished, and hair parted as nicely as if he were dressed by his valet in his New York apartments. How did he do it? We never knew. He kept no servant; he took his regular turn in the ditches, in the mud, or torrid sun, or smothering rain. No night alarm came that did not find Kane first to spring to the trench— and yet he did it, somehow. The courteous phrases of politest speech fell ever from his ready lips, as easily as they would have done in the *boudoir* of any belle in the metropolis.

The shrieking of a shell or tingling hiss of a sharpshooter's close-aimed bullet never came so near as to interrupt whatever polished expression of thanks, regret, or comment he might be

RELICS OF THE BATTLE

uttering. And it was the real thing, too. The gentle heart was there. No man was readier to bind a wound or aid a sun-struck soldier in the ranks; none more ready to deny himself a comfort or a luxury to help a more needy comrade. A braver man, a surer or more reliable officer, never trod in shoe-leather. A grand example to our pessimistic, socialistic friends and cheap demagogues of the sterling worth and noble, chivalric character of a "society man of wealth." He is a living type of *"Bel a faire peur,"* without the idiotic sentimentality of that maudlin hero, and with all his other characteristics.

Greenway and Tiffany. The one a Harvard football-player, just out, plunging into the great game of war with all the zest he formerly found in the great college game. The other the petted son of wealthy parents, also a college graduate, and the idolized fiancé of his childhood's sweetheart. Equally ready for fight or fun, they were the finest type of youthful manhood to be found. Endowed by Nature with every gift, educated at the best of colleges, bred in the best of society, ready to enter upon the most desirable of careers, they threw all upon the altar of country's love. They entered battle as one might go to a game or begin a play. All of unbounded zeal, youthful enthusiasm, restless energy, keen enjoyment—everything seemed to be equally acceptable to them, and no discomfort ever assumed any guise other than that of a novel and untried sensation.

They are the type of our young manhood—our representative American youth—as Roosevelt is of its vigorous manhood. They are the salt of the earth, and Kane—is both salt and spice. All were comrades in arms, types of American manhood unspoiled by Fortune's favours, capable of anything and everything. Such men mould the destiny of this great nation, and in their hands it is safe.

But neither of these two regiments is a fair type of the Volunteers; they are the two extremes. For a type, take the 1st Illinois. They were a Chicago regiment with fifteen years' service, and they enlisted in a body to a man. They reached the firing line on the 10th and participated in the fight with two battalions, with distinguished gallantry. The third battalion was detailed on the necessary but unpleasant duty of caring for the yellow fever hospital at Siboney. These city-bred Volunteers peeled off their coats, buried yellow fever corpses, policed the hospital and hospital grounds, and nursed the victims of the scourge. They did not utter a complaint nor ask for a "soft" detail; they did their duty as they found it. Another battalion was detailed immediately after the surrender to guard the Spanish prisoners.

This most thankless duty was performed by them with fidelity and care. The commander of the battalion and half his officers were proficient in the Spanish language as a part of their preparation for the campaign, and they soon established cordial relations with the prisoners they were set to guard. It was a trying duty, but they performed it faithfully. Sickness visited this battalion, and sometimes guard duty had to be performed with only one day off, but they never whimpered.

The other battalion was detailed after the surrender to do stevedore work at the commissary depot. The slender clerks and soft-handed city men slung boxes of hardtack and sacks of bacon and barrels of coffee, and performed manual labour with all the faithfulness that would be expected of men accustomed to such work, and with never a complaint. The sanitary measures of this regiment were perfect, and they bore themselves like Regulars. It is now recognized that this is a compliment to any Volunteer organization.

Chapter 10

Sufferings of the 5th Army Corps

In such a campaign as that of Santiago, a certain amount of suffering is inevitable. In such a climate as that of southern Cuba, a certain amount of disease is unavoidable. In the very hot-bed of yellow fever and malaria, no army could hope to escape without contracting these diseases; and in a campaign conducted with the marvellous celerity of the one at Santiago, some difficulty in forwarding supplies must necessarily be encountered.

The root of all our difficulties lay in the fact that the commanding general had under him supply departments whose officers reported to heads of bureaus not under control of the corps commander. This caused unnecessary delays in obtaining supplies, entailed confusion in their distribution, and led to suffering beyond what was necessarily the result of the climate and the campaign.

A brief description of the method of obtaining supplies will make this point more clear. When a given article was wanted, whether it was soap, quinine, tentage, or transportation, a requisition upon the chief of the proper bureau at Washington had to be made, with full statement of the reasons for the request; this requisition had to be approved by all intermediate commanders and go through military channels to the chief of the bureau, who might or might not be convinced of the necessity for the article wanted. His action being endorsed thereon, the requisition returned through the same devious route, and possibly might be followed in course of time, either by invoices from some distant purchasing agent of the required articles, or by directions of the bureau chief to make further explanations. The usual length of time allowed for an official communication through military channels, in time of peace at home, from any regimental headquarters to Washington and return, is from ten to thirty days. Here was the first cause of suffering.

If the heads of the supply departments in the field, beginning at Tampa, could have acted promptly upon the orders of their respective commanding officers, without the action of any other authority, unnecessary delay would have been avoided.

To illustrate this point: The Gatling Gun Detachment was ordered to be equipped with revolvers upon reporting to the detachment commander, and this order was issued on the 11th of June, before sailing from Port Tampa. They did not so report, and it devolved upon the detachment commander to make requisition for the necessary equipment. This was done, but no revolvers arrived. The invoices for revolvers reached the detachment commander on the 15th of September, at Fort Leavenworth, Kansas, where he was then, on leave of absence, sick, ten days after the detachment was disbanded.

This is an extreme case, but the same difficulty was experienced in obtaining supplies of all descriptions. It was, therefore, very difficult for a quartermaster, a commissary, a medical officer, or any other officer whose duty it was to obtain supplies, to have the same when emergency demanded it. The necessity for supplies could not always be foreseen, the quantity desired could not always be estimated for with precision, and it followed that sometimes there was a deficiency when the articles were needed.

Again, the transportation of the 5th Army Corps could not be made available at first to carry supplies up from the landing-place. The troops had drawn travel rations, which lasted them until they disembarked. The first supply problem, upon landing, was that of issuing rations; and, at the moment when every available boat was engaged in carrying troops ashore, it became necessary to put rations ashore also. The exigency demanded the speedy disembarkation of the greatest possible number of men. The fight of La Guasimas emphasized the necessity of get-

Cieba tree under which General Toral surrendered

ting men to the front. It was no time to delay the movement of troops for the purpose of waiting on wagons, tentage, or rations. The safety of the expedition, the fate of the whole campaign, depended upon energetic and rapid movement to the front. Consequently regiments were put forth with only such amounts of rations and tentage as they could carry upon their backs. It will be readily seen that this amount was very limited, and the only tentage possible was the shelter tent.

There were 118 wagons in the hold of the *Cherokee*, but it was not practicable to delay the disembarkation of the corps and hazard the fate of the whole campaign by utilizing the only wharf and all the boats two or three days to land these wagons. By the time they could be taken off, the rains had made the roads almost impassable, and they could not all be used. It was therefore a daily struggle to get enough rations forward to feed the fighting-line from day to day. Greatly to the credit of those who performed the duty, it can be said that, with rare exceptions, all the soldiers of the 5th Army Corps had every day, when they could possibly cook the same, hardtack and bacon, roast beef, and coffee. This much was accomplished in the face of insurmountable obstacles by the heroic exertions of the pack-train.

When the 1st of July arrived, and the battle began, it was ordered that all soldiers carry three days' rations. The heat was intense, the fight exceedingly hot, and marching through the jungle extremely difficult. The consequence was that the soldiers threw aside all impedimenta in order to fight more effectively, and, of course, the rations went with the blankets and the overcoats. The man who held on to a canteen and haversack was fortunate; very many abandoned the haversack, and a considerable number abandoned everything except rifle and ammunition. That was what won the fight; but it made hungry men, and it caused men to sleep on the wet ground under the open sky, without blankets or tentage. The pack-train continued its magnificent work. During the fighting it had to bring ammunition. The men were supposed to have three days' rations. As soon as the deficiency became known to the higher officials, the pack-train began to bring food. Commissary depots were established immediately in rear of the firing-line, and issues of hardtack, bacon, and coffee, which were about the only components of the ration that could be brought forward in sufficient quantities, were made without formality or red tape. It was almost impossible to get a sufficient quantity of even these components to the front. Sometimes the ration was a little short. Bacon and hardtack for seventeen consecutive days, after three weeks of travel ration, do not form the most appetizing diet in the world. The exposure consequent upon the fighting and lack of tentage had its inevitable result in sickness.

The same difficulties which had beset the quartermaster and commissary departments were also encountered by the surgeons. Hospital accommodations were scanty, the quantity of medicines available was very limited, the number of wounded men disproportionately large, and, when sickness was added to the wounds, the small number of surgeons available at the front were not able to give the individual attention and scientific treatment which forms a part of our admirable medical system in time of peace. There were only three or four ambulances available until after the 11th of July. A considerable number of the surgeons were on duty at the general hospitals far in the rear; the number at the front was not sufficient to attend to all the duties which devolved upon them. This deplorable condition reacted, causing a greater amount of illness. To add to this difficulty, the Volunteers began to suffer excessively from the results of their own ignorance and carelessness; and when the yellow fever scourge was added to all the other difficulties which beset the 5th Corps, the outlook became gloomy.

The attempt has been made in the foregoing exposition of the conditions at Santiago to represent fairly the difficulties under which all parts of the army laboured. The fact remains, nevertheless, that there was an appalling amount of suffering due to causes which might have been foreseen and which were easily preventable.

On the 18th day of July the transports entered the harbour of Santiago. From that day forward there was unlimited wharfage at disposal, and there were excellent macadamized roads leading to all parts of the command. The fall of Santiago had been

UNDERGROWTH IN CUBA

foreseen more than a week, and if there was not a sufficient quantity of wagons present on board the ships, there had been ample time to make telegraphic requisition for them to Washington. Up to the surrender, the suffering from sickness had been exceedingly light. There was something stimulating about the nervous strain and excitement of the time which kept the men up to their work; but the inadequacy of the medical supplies on hand had been amply demonstrated by the 10th. and it had become fully apparent that the medical corps was unable to handle the number of patients on hand. The previous remark about the practicability of telegraphing to headquarters for additional force applies to this department also.

The principal sufferings after the surrender were due to four causes: first, improper clothes; second, improper food; third, lack of shelter; fourth, lack of proper medical attention.

In regard to clothing and these other necessaries, it should be borne in mind that the corps which went to Santiago was virtually the Regular Army. Every regiment which went to Tampa went there ready for service. Its equipment was just as complete on the 26th of April as it was on the 6th of June. There should have been no problems to solve in regard to them—and yet there were many.

First—Clothing.

The troops wore the same clothing to Cuba they had brought from Sheridan, Assinniboine, and Sherman. They wore winter clothing for their service in the torrid zone, and those who received summer clothing at all received it late in August, just in time to return to the bracing breezes of Montauk Point, where, in their enfeebled condition, winter clothing would have been more suitable. It did not require a professor of hygiene to foresee that the winter clothing used in northern Michigan would not be suitable for campaigning in southern Cuba in July; or that summer clothing suitable for southern Cuba would be too light for men returning to the northern part of Long Island. Is it to be concluded that it was impossible to obtain summer clothing for 18,000 men between the 26th of April and the 6th of June?

SECOND—IMPROPER FOOD.

Most of the troops were embarked upon the transports by the 10th of June. Their food on transport consisted of the travel ration: canned roast beef, canned baked beans, canned tomatoes, and hardtack, with coffee, were the components. They subsisted upon this food, imprisoned in fetid holds of foul transports, unfit for the proper transportation of convicts, until the 25th day of June, when they disembarked. On drawing rations for the field it was found that the field ration would be of the same components, with the addition of bacon and minus the baked beans and tomatoes. During the emergency, up to include the 18th day of July, this was the ration. Occasionally a few cans of tomatoes found their way to camp, but rarely. The ration was always short, such as it was, but this the soldiers could have endured and did endure without a murmur.

But on the 18th of July, with unlimited wharfage at a distance of two miles and a half, with excellent roads, and with abundance of transportation (see Gen. Shafter's Official Report), and with surrender foreknown for a sufficient length of time to have brought any quantity of vegetables from New York City, the ration continued to be bacon, canned beef, hardtack, and coffee. Finally, about the 25th of July, small amounts of soft bread began to be doled out, and an occasional issue of frozen fresh beef was made. It was soon demonstrated that not sufficient fresh beef could be made available. The vegetables which had been brought had nearly all spoiled on the transports. Hundreds of barrels of potatoes and onions were unloaded upon the docks and were so badly decayed as to make them useless. These vegetables had been drifting about the Caribbean Sea and upon the Atlantic Ocean since the 9th and 10th of June. Occasionally it was practicable to get a quarter or a half ration of potatoes and half of the usual allowance of canned tomatoes, but that was all.

It did not require a professor of hygienic dietetics to predict that men fed in the tropics upon a diet suited to the icy shores of Greenland would become ill, especially when they were clad in a manner suited to the climate of Labrador. Are we to

CUBAN RESIDENCE

conclude that it was impossible to get rice, beans, canned fruits, canned corn, and other vegetables to take the place of potatoes and onions?

THIRD—LACK OF SHELTER.

The allowance of tentage was prescribed for each regiment. Granted that it was impossible to get tentage up until after the surrender; yet it should have been practicable to forward tentage over two and one-half miles of macadamized roads. Yet whole regiments remained without tentage until they embarked for the United States. The 13th Infantry did not get tentage until the 5th of August. The 20th Infantry and the 3rd Infantry obtained a portion of their tentage about the same time, but a large part of these regiments remained under shelter tents until they re-embarked. The 1st Illinois and the 34th Michigan remained in shelter tents until the 15th of August, at which time the author embarked for the United States. These regiments are fair examples.

The Gatling Gun Detachment was provided with shelter-halves and remained under them until the 10th of August. Repeated applications for proper tentage were made, accompanied by medical certificates that the issue of tentage was imperatively necessary for the health of the command. Endorsements thereon by the chief quartermaster of the 5th Corps as late as the 5th of August show that there was no available tentage for issue. Application was made to the regimental commander, 13th Infantry, for a portion of regimental tentage for the detachment of the 13th Infantry; but, in spite of the fact that the reduced regiment had on hand all the canvas prescribed for the full regiment, none could be obtained for the detachment. The detachment commander was entirely without tentage from the 25th of June until the 5th of August—forty-five days in the rainy season in Cuba, exposed to the torrid sun by day, to chilling dews by night, and the drenching rains of the afternoon, without shelter from any inclemencies of the weather, and this in spite of repeated applications to proper authorities for the suitable allowance of tentage. Is it any wonder that men grew sick, and that death stalked broadcast through the camp of the 5th Corps, under these conditions?

FOURTH—LACK OF PROPER MEDICAL ATTENDANCE.

The surgeons who were at the front with the firing-line worked heroically, but were burdened beyond their physical powers. Owing to the foregoing causes, great numbers of men became ill as soon as the strain and tension of the battle were relieved. It was not uncommon to find twenty or twenty-five per cent of a command on the sick-report, and in some cases the sick-list went as high as fifty per cent. There were no well men in the 5th Army Corps. Those who refused to go on the sick-report were, nevertheless, sick. The author has yet to find a single member of the expedition who did not suffer from the climatic fever. The surgeons themselves were not exempt, and the very limited supply of doctors was speedily decreased by sickness. Were there no doctors in the United States who were willing to come to Cuba?

Up to the 25th of July the supply of medicines was very deficient. There was never a sufficient supply of ambulances. The accommodations in the hospitals were even worse than those on the firing-line. A sick soldier on the firing-line could always find some comrade who would cut green boughs or gather grass for a bed, but the one who went to the hospital had to lie on the ground. The supply of hospital cots was ridiculously inadequate, and this condition did not improve.

The difficulty of obtaining adequate medical attendance may be illustrated by the case of Priv. Fred C. Elkins, of the 17th Infantry, member of the Gatling Gun Detachment. Priv. Elkins had been hurt in the fight on the 1st of July and had been sent to the hospital. He found the accommodations so wretched that he feigned improvement and returned to his detachment. He remained with the detachment until the 14th of July, improving so far as his injury was concerned, but contracted the climatic fever. During this time he was prescribed for twice by the assistant surgeon with the Rough Riders, Dr. Thorpe, previous to the time this regiment was moved westward on the firing-line. His condition became worse, and on the 12th of July Dr. Brewer, 1st lieutenant and assistant surgeon with the 10th Cavalry, was called upon to examine him. This surgeon had then under treatment over 100 cases pertaining to his proper command, and was himself ill, but he readily came and inspected the patient. He promised to send medicines for him, but in the rush of overwork forgot to do so, and on the 13th of July he was again summoned. This time he sent a hospital attendant to take the patient's temperature, which was 104 degrees. No medicines were sent.

On the 14th of July the patient became delirious. The detachment commander went in person to request the same surgeon to attend to the case, he being the only one available at that time. The hospital attendant was again ordered to take the temperature. At the end of an hour even this had been neglected. The hospital man was sick, and had been without sleep for fifty hours. Priv. Elkins was put upon a board and carried to Brewer's tent, with his descriptive list in his pocket. The surgeon was told the name of the patient and the facts that he was related to a distinguished family and had been recommended for a commission for gallantry upon the field of battle. Dr. Brewer was himself suffering at the time, with a temperature of 103 degrees, but he rose from his own sick-bed and administered remedies which relieved the patient.

The following day, the third of his illness, Dr. Brewer was found to be suffering from yellow fever, and was carried back to the yellow fever hospital at Siboney along with Priv. Elkins. He had been sick all the time, but had done his best. Priv. Elkins improved sufficiently to write a letter to his commanding officer from the hospital at Siboney, on the 25th of July, which reached that officer at Fort Leavenworth, Kansas, on the 12th day of September. In spite of the fact that the patient was furnished with descriptive list, and was specially commended to the care of the surgeon as a soldier marked for extreme gallantry, all trace of him had been lost; and although two private detectives were searching for him a month, no further clew had been found to his whereabouts or fate as late as the 1st of October. Even if his descriptive list had not been furnished with this man, the fact that he was alive and rational enough on the 25th day of July to write a letter concerning his approaching discharge should have made it easy for some record of his case to have been kept.

But this one isolated case sinks into insignificance beside the condition in which some of the sick were left by commands returning to the United States. All cases of yellow fever suspects were left behind, and in the mad scramble to embark for the return voyage many of these were left without proper attention or supplies.

Gen. Kent's Division had left by the 11th of August. The following extract from a letter dated Santiago de Cuba, August 12, 1898, will convey some idea of the condition in which the sick of this division were left:

Yesterday Gen. Kent's Division left for Montauk, and they left behind 350 sick, many of them too ill to care for themselves. This humane country, of course, left ample care for them? There was left one surgeon, one steward, and one case of medicines. Many of these men are too ill to rise. They are 'suspected' of having yellow fever. They are suffering from Cuban malaria, and many of them from diarrhoea. There was not left a single bed-pan for this battalion of bed-ridden, suffering humanity, nor any well men to nurse the sick. There was not even left any to cook food for them. Those left by the 9th Infantry had to bribe marauding, pilfering Cubans, with a part of their rations, to carry food to the camp of the 13th, where there were a few less ill, to get it cooked.

They are too sick to dig sinks; some are delirious. When the poor emaciated wrecks of manhood have to obey the calls of Nature, they must either wallow in their own filth or stagger a few paces from their wet beds on the slimy soil to deposit more germs of disease and death on the surface already reeking with ghastly, joint-racking rheums.

There were left less than fifty cots for these 350 sick men— men compelled by sheer weakness to lie on the ground which will soon lie on them, if enough strong men are left by that time to cover them mercifully over with the loath-

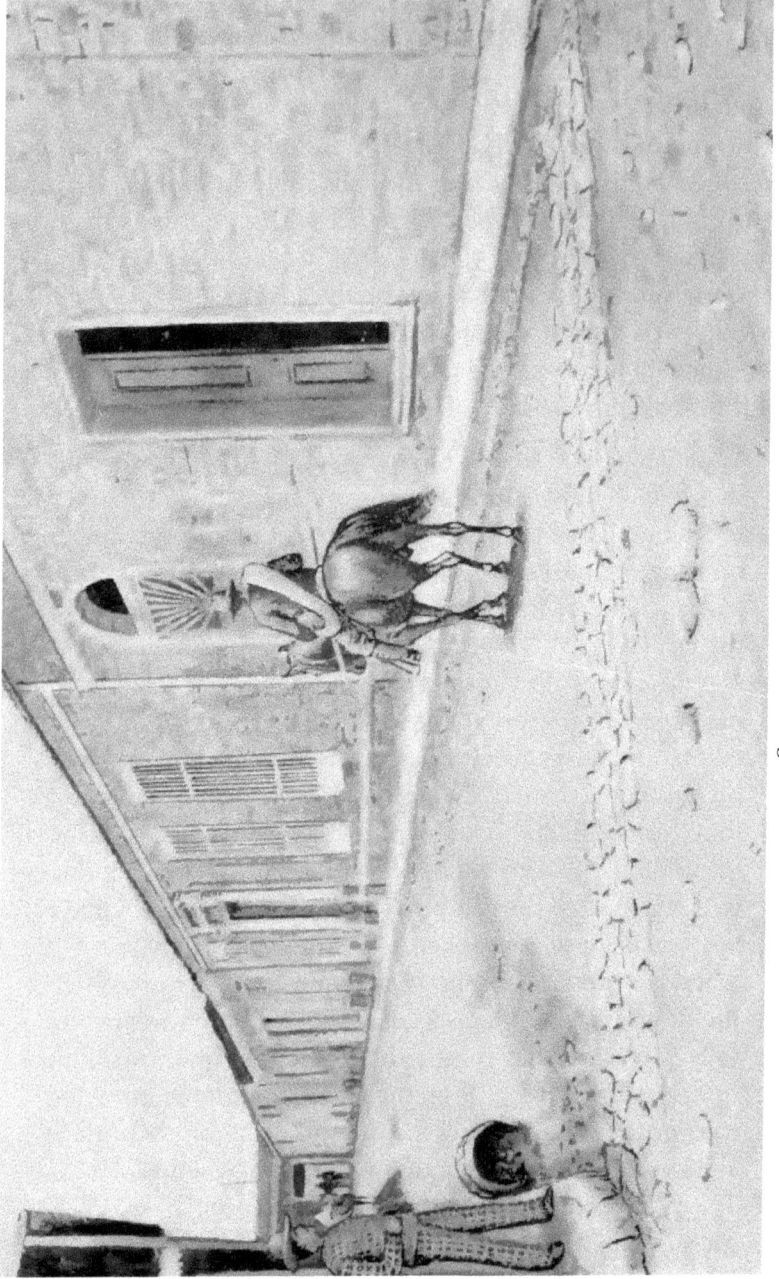

Santiago street scene

some, reeking vegetable detritus which passes here for soil, and which is so fairly animate that you can see every spadeful of it writhe and wriggle as you throw it over the rotting hour-dead shell of what was a free American citizen and a Chevalier Bayard.

"When the last man and wagon of the flying division disappeared over the hill toward health and home, a despairing wail went up from the doomed 350 left in this condition of indescribable horror. 'We are abandoned to die!' they cried; 'we are deserted by our own comrades in the hour of danger and left to helplessly perish!'

These men are those who fought the climate, hunger, and the enemy on the battle-field which has shed so much undying glory on the American arms. They are the men who have accomplished unheard-of feats of endurance and performed incredible feats of valour on the same ground—not for Cuba, but at the call of duty. They are citizens. They are brave soldiers who have done their full duty because it was duty.

The mail facilities were wretched. Cords of mail were stacked up at Siboney for weeks; and although there was more transportation on hand than could be used, the officer detailed to attend to the mail business of the corps, Lieut. Saville, of the 10th Infantry, could not succeed in securing a wagon to haul this mail to the front. Since the corps returned to the United States a dozen letters have reached the author which have chased him by way of Santiago and Montauk, since dates between the 1st and 20th of July, inclusive. The person to whom these letters were addressed was well known to every officer and employee in the corps, and if the mail addressed to one so well known could go astray in this manner, what could an unknown private expect? This may seem like a little hardship, but to men in the weakened and enfeebled condition of the survivors of the 5th Corps a letter from home was both food and medicine. Scores of men who are to-day rotting in Cuban graves died of nostalgia, and might have lived if they had received the letters from home which were sent to them.

CHAPTER 11

The Cause

The causes of these conditions are not far to seek. The United States has not had an army since 1866. There has been no such a thing as a brigade, a division, or a corps. There has been no opportunity to study and practice on a large scale, in a practical way, the problems of organization and supply. The Army has been administered as a unit, and the usual routine of business gradually became such that not a wheel could be turned nor a nail driven in any of the supply departments without express permission, previously obtained from the bureau chief in Washington. The same remarks apply equally to all the other staff departments. The administration had become a bureaucracy because the whole Army for thirty years had been administered as one body, without the subdivisions into organizations which are inevitable in war-time and in larger bodies.

War became a reality with great suddenness. Those who have grown grey in the service, and whose capacity, honesty, and industry had never been and can not be impeached, found themselves confronted with the problem of handling nearly three hundred thousand men, without authority to change the system of supply and transportation. The minutest acts of officers of these departments are regulated by laws of Congress, enacted with a view of the small regular force in time of peace, and with no provisions for modifications in war. In authorizing the formation of large volunteer armies, Congress did not authorize any change in the system of administration or make any emergency provision. As before, every detail of supply and transportation had to be authorized from the central head.

The administrative bureaus were handicapped to some extent by incompetent and ignorant members. Late in the campaign it was learned that the way to a "soft snap" was through the Capitol, and some came in that way who would certainly never have entered the Army in any other.

There were alleged staff officers who had tried to enter the service through the regular channels and who had failed, either by lack of ability or bad conduct, to keep up with the pace set by classmates at the Academy; there were others who were known as failures in civil life and as the "black sheep" of eminent families; and there were some who must have been utterly unknown before the war, as they will be afterward.

How these persons ever obtained places high above deserving officers of capacity and experience is a question which cries aloud for exposure—but in a good many cases they did. Indeed, it is to be observed that, for that matter, the next register of the Army will show a great many more promotions into the Volunteer service, of officers who never heard a hostile bullet during the war, who never left the United States at all, than it will of deserving officers who bore the heat and burden of the march and the battle.

The most discouraging thing about it all to a line officer is that this same register will afford no means of determining who did the service and who did the "baby act." Lieut. Blank will be borne thereon as major and subsequently colonel of the Steenth Volunteers (which never left the State rendezvous, probably) during the war with Spain; Lieut. Blank No. 2 will be carried on the same book as second lieutenant, —— Infantry, during the same war. The gentle reader will at once "spot" the man who was so highly promoted as a gallant fellow who distinguished himself upon the bloody field; the other will be set down as the man who did nothing and deserved nothing.

Yet—the ones who went received no promotion, and those who staid behind and by their careless incompetence permitted camps amid the peaceful scenes of homes and plenty to become the hot-beds of fever and disease—these are the ones borne as field and other officers of the Volunteers.

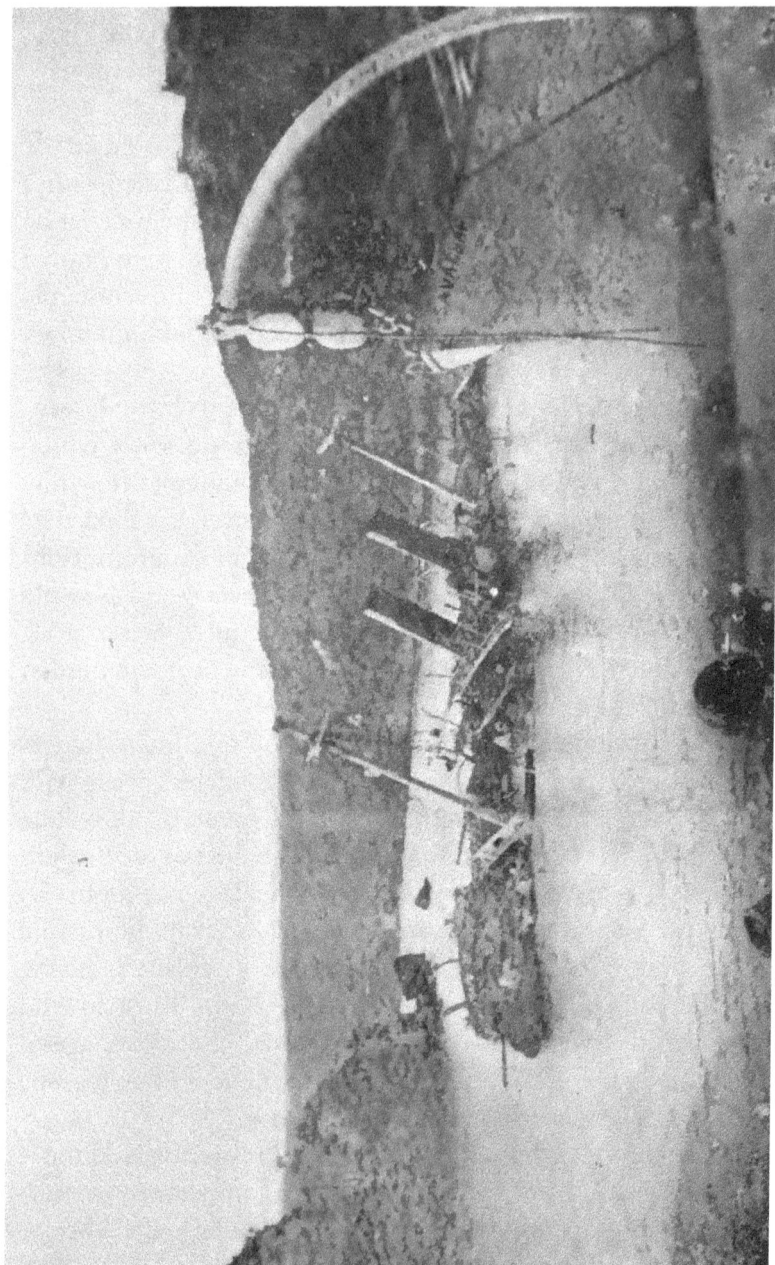

REINA MERCEDES SUNK BY THE *IOWA* NEAR THE MOUTH OF SANTIAGO HARBOUR

To illustrate some of the material with big titles sent to "assist" in running the staff departments, two incidents will suffice.

On the 11th of June, at a certain headquarters, it was desired to send a message, demanding reply, to each transport. A grey-haired officer turned to another and said, "Whom shall we send with this? Will So-and-so do?" naming one of the before-mentioned civil appointments.

"For heaven's sake, no! He would tie up the whole business. Send an orderly," was the reply.

The orderly, an enlisted man of the Regulars, was sent. The officer thus adjudged less competent to carry a message than a private soldier was perhaps actuated by a high sense of duty; but he filled a place which should have been occupied by an experienced and able officer—no, he did not fill it, but he prevented such a man from doing so.

The second incident was related by an officer on a transport bound for home. Say his name was—oh well, Smith.

Smith went, on the 20th of July, to a certain headquarters in the field on business. Those who could have attended to it were absent, but there was one of the recent arrivals, a high-ranking aide, there, and he, sorry for Smith's worn-out look of hunger, heat, and thirst, asked if he would have a drink. Smith, expecting at the best a canteen of San Juan River water, said he was a little dry.

The newly-arrived clapped his hands, and, at the summons, a coloured waiter in spotless white duck appeared.

"Waitah, take this gentleman's ordah," said the host.

Smith, greatly astonished, asked what could be had, and was yet more astonished to learn that he could be served with Canadian or domestic whisky, claret, champagne, or sherry. Much bewildered, and utterly forgetting the awful dangers of liquor in the tropics, he called for Canadian Club. When it came, on a napkin-covered tray, he looked for water, and was about to use some from a bucket full of ice which he at that moment espied.

"Aw! hold on," exclaimed the host; "we nevah use that, don't y' know, except to cool the *apollinaris*. Waitah, bring the gentleman a bottle of *apollinaris* to wash down his liquor."

Within half a mile were soldiers and officers lying sick in hospital on the ground, eating hardtack and bacon, and drinking San Juan straight, because hospital supplies and rations could not be got to the front!

It was this same officer who explained that he approached his headquarters "by rushes," upon his arrival, for fear the enemy would see him and consider this reinforcement a violation of the truce.

These are two examples of some of the able assistants from civil life who were sent to help feed, clothe, and transport the 5th Corps.

With such assistants, is it any wonder that, under such extraordinary circumstances as those encountered in Cuba, a system designed for peace and 25,000 men weakened in some respects when the attempt was made to apply it to 300,000 in time of war?

The great wonder is that it did the work as well as it did. And this was due to the superhuman exertions of the chief officers of the supply departments and their experienced assistants. These men knew no rest. They were untiring and zealous. On their own responsibility they cut the red tape to the very smallest limit. Instead of the regular returns and requisitions, the merest form of lead-pencil memorandum was sufficient to obtain the necessary supplies, whenever they were available. This much was absolutely necessary, for these officers were personally responsible for every dollar's worth of supplies and had to protect themselves in some degree. As it is now, many of them will find it years before their accounts are finally settled, unless some provision be made by law for their relief. This disregard of routine was essential; but how much to be desired is a system suited to the exigencies of the service, both in peace and war!

There is a lesson to be learned from these experiences, and it is this: The commanding officer of any army organization should not be hampered in the matter of supplies by having to obtain the approval or disapproval of a junior in rank, in a distant bureau, who knows nothing about the circumstances. In other words, the system which causes the staff departments of the United States Army to regard a civilian as their head, and makes them virtually independent of their line commanders, is an utterly vicious system. If an officer is competent to command an organization, he should be considered competent to look after the details of its administration, and should be held responsible, not only for its serviceable condition at all times, but for the care of its property and for all the other details connected with its service.

The quartermaster, or commissary, or other officer of a supply department should not know any authority on earth higher or other than the officer in command of the force he is to serve, except those in the line above such chief, and then only when such orders come through his chief.

The commanding officer having ordered supplies to be procured, there should be no question whatever in regard to their being furnished. They should come at once and without fail. If they were not necessary, hold him responsible.

This theory of administration eliminates the bureaucracy which has insidiously crept upon the Army, and relegates to their proper position the supply departments.

The General Staff proper has a higher field of usefulness than the mere problems of supply. Its business is to care for the organization, mobilization, and strategic disposition of all the forces, both naval and military, of the United States. Its head should be the President, and the two divisions should be under the general commanding the Army and the admiral commanding the Navy. The remainder of this staff should be composed of a small but select personnel, and should limit its duties exclusively to those set forth above.

CHAPTER 12

The End of the Gatling Gun Detachment

The detachment received permission on the 10th of August to use any standing tentage which it could find, and it was thoroughly under shelter an hour after this permission was received. The climate of Cuba was not so disagreeable when one could look at it through the door of a tent, but we were not destined to enjoy our tentage very long. On the 15th, at two o'clock, orders were received to go on board the Leona at Santiago, bound for Montauk Point, and at half-past five o'clock men, guns, and equipment were duly stowed for the voyage home.

It was much more agreeable than the one to Cuba, The transport was not crowded, the men had excellent hammocks, which could be rolled up during the day, thus leaving the whole berth deck for exercise and ventilation, and the Leona was a much better vessel than the *Cherokee*.

The detachment finally disembarked at Montauk Point on the 23rd, passed through the usual detention camp, and was assigned a camping-place. It was disbanded per instructions from headquarters, Montauk Point, on the 5th of September, the members of the detachment returning to their respective regiments, well satisfied with the work they had done and with each other.

In concluding this memoir the author desires to pay a personal tribute of admiration and respect to the brave men composing the detachment, both individually and collectively. Some of them have figured more prominently in these pages than others, but there was not a man in the detachment who was not worthy to be called the highest term that can be applied to any man—a brave American soldier.

Appendices

Appendix 1

Headquarters U. S. Troops
Santiago de Cuba
July 19, 1898

GENERAL ORDERS No. 26.

The successful accomplishment of the campaign against Santiago de Cuba, resulting in its downfall and surrender of Spanish forces, the capture of large military stores, together with the destruction of the entire Spanish fleet in the harbour, which, upon the investment of the city, was forced to leave, is one of which the Army can well be proud.

This has been accomplished through the heroic deeds of the Army and its officers and men. The major-general commanding offers his sincere thanks for their endurance of hardships heretofore unknown in the American Army.

The work you have accomplished may well appeal to the pride of your countrymen and has been rivalled upon but few occasions in the world's history. Landing upon an unknown coast, you faced dangers in disembarking and overcame obstacles that even in looking back upon seem insurmountable. Seizing, with the assistance of the Navy, the towns of Baiquiri and Siboney, you pushed boldly forth, gallantly driving back the enemy's outposts in the vicinity of La Guasimas, and completed the concentration of the army near Sevilla, within sight of the Spanish stronghold at Santiago de Cuba. The outlook from Sevilla was one that might have appalled the stoutest heart. Behind you ran a narrow road made well-nigh impassable by rains, while to the front you looked upon high foot-hills covered with a dense tropical growth, which could only be traversed by bridle-paths terminating within range of the enemy's guns. Nothing daunted, you responded eagerly to the order to close upon the foe, and, attacking at El Caney and San Juan, drove him from work to work until he took refuge within his last and strongest entrenchment immediately surrounding the city. Despite the fierce glare of a Southern sun and rains that fell in torrents, you valiantly withstood his attempts to drive you from the position your valour had won, holding in your vice-like grip the army opposed to you. After seventeen days of battle and siege, you were rewarded by the surrender of nearly 24,000 prisoners, 12,000 being those in your immediate front, the others scattered in the various towns of eastern Cuba, freeing completely the eastern part of the island from Spanish troops.

This was not done without great sacrifices. The death of 230 gallant soldiers and the wounding of 1,284 others shows but too plainly the fierce contest in which you were engaged. The few reported missing are undoubtedly among the dead, as no prisoners were taken. For those who have fallen in battle, with you the commanding general sorrows, and with you will ever cherish their memory. Their devotion to duty sets a high example of courage and patriotism to our fellow-countrymen. All who have participated in the campaign, battle, and siege of Santiago de Cuba will recall with pride the grand deeds accomplished, and will hold one another dear for having shared great suffering, hardships, and triumphs together.

All may well feel proud to inscribe on their banners the name of Santiago de Cuba.

By command of Major-General Shafter.

Official: *John B. Miley*

E. J. McClernand, Aide. Asst. Adj.-Gen.

Appendix 2

REPORT OF
MAJOR-GENERAL WM. R. SHAFTER
COMMANDING

September 13, 1898
Sir,

I have the honour to submit the following report of the campaign which terminated in the fall of Santiago de Cuba and the adjacent territory, and the establishment of the military government therein.

The expedition was undertaken in compliance with telegraphic instructions of May 30, 1898, from Headquarters of the Army, in which it was stated:

> Admiral Schley reports that two cruisers and two torpedo boats have been seen in the harbour of Santiago. Go with your force to capture garrison at Santiago and assist in capturing harbour and fleet.

On this date there were a large number of transports in Port Tampa Bay, which had been collected for the purpose of an expedition which it had been previously contemplated I should command, and for such other emergencies as might arise. Orders were immediately given for loading aboard those transports the necessary subsistence and quartermaster supplies, and for the embarkation of the authorized number of troops and their material. General Orders No. 5, from these headquarters, indicate the organizations it was at first proposed to take.

The order is as follows:

Headquarters 5th Army Corps
Tampa, Fla.
May 31, 1898

G. O. 5.

The following troops will hold themselves in readiness to move immediately on board transports upon notification from these headquarters:

1. The 5th Army Corps.
2. The Battalion of Engineers.
3. The detachment of the Signal Corps.
4. Five squadrons of cavalry, to be selected by the commanding general of the cavalry division, in accordance with instructions previously given.
5. Four batteries of light artillery, to be commanded by a major, to be selected by the commanding officer of the light artillery brigade.
6. Two batteries of heavy artillery, to be selected by the commanding officer of the siege artillery battalion, with eight (8) guns and eight (8) field mortars.
7. The Battalion of Engineers, the infantry and cavalry will be supplied with 500 rounds of ammunition per man.
8. All troops will carry, in addition to the fourteen (14) days' field rations now on hand, ten (10) days' travel rations.
9. The minimum allowance of tentage and baggage as prescribed in G. O. 54, A. G. O., c. s., will be taken.
10. In addition to the rations specified in paragraph 8 of this order, the chief commissary will provide sixty (60) days' field rations for the entire command.
11. All recruits and extra baggage, the latter to be stored, carefully piled and covered, will be left in camp in charge of a commissioned officer, to be selected by the regimental commander. Where there are no recruits available, the necessary guard only will be left.

12. Travel rations will be drawn at once by the several commands, as indicated in paragraph 8.

By command of Maj.-Gen. Shafter.

E. J. McClernand

A. A. G.

This order was afterwards changed to include twelve squadrons of cavalry, all of which were dismounted because of lack of transportation for the animals, and because it was believed, from the best sources of information obtainable, that mounted cavalry could not operate efficiently in the neighbourhood of Santiago. This was found subsequently to be correct.

The facilities at Tampa and Port Tampa for embarking the troops and the large amount of supplies required were inadequate, and with the utmost effort it was not possible to accomplish this work as quickly as I hoped and desired.

On the evening of June 7th I received orders to sail without delay, but not with less than 10,000 men.

The orders referred to caused one division, composed of Volunteer troops, commanded by Brig.-Gen. Snyder, and which it had been intended to include in my command, to be left behind. I was joined, however, by Brig.-Gen. Bates, who had already arrived on transports from Mobile, Ala., with the 3rd and 20th Infantry and one squadron of the 2nd Cavalry with their horses, the latter being the only mounted troops in my command.

After some of them had already reached the lower bay, telegraphic instructions were received from the honourable Secretary of War, directing that the sailing of the expedition be delayed, waiting further orders. This delay was occasioned by the Navy reporting that a Spanish war vessel had been sighted in the Nicholas Channel. The ships in the lower bay were immediately recalled. On the next day, in compliance with instructions from the adjutant-general of the Army, the necessary steps were taken to increase the command to the full capacity of the transports, and the expedition sailed on June 14th with 815 officers and 16,072 enlisted men.

The passage to Santiago was generally smooth and uneventful. The health of the command remained remarkably good, notwithstanding the fact that the conveniences on many of the transports, in the nature of sleeping accommodations, space for exercise, closet accommodations, etc., were not all that could have been desired. While commenting upon this subject, it is appropriate to add that the opinion was general throughout the Army that the travel ration should include tomatoes, beginning with the first day, and that a small quantity of canned fruit would prove to be a most welcome addition while travelling at sea in the tropics. If the future policy of our Government requires much transportation for the military forces by sea, definite arrangements should be determined upon to provide the necessary hammock accommodations for sleeping. Hammocks interfere immeasurably less than bunks with the proper ventilation of the ships and during the day can be easily removed, thus greatly increasing space for exercise; moreover, they greatly diminish the danger of fire.

While passing along the north coast of Cuba one of the two barges we had in tow broke away during the night, and was not recovered. This loss proved to be very serious, for it delayed and embarrassed the disembarkation of the army. On the morning of June 20th we arrived off Guantanamo Bay, and about noon reached the vicinity of Santiago, where Admiral Sampson came on board my headquarters transport. It was arranged between us to visit in the afternoon the Cuban general (Garcia) at Aserraderos, about eighteen miles to the west of the Morro. During the interview Gen. Garcia offered the services of his troops, comprising about 4,000 men in the vicinity of Aserraderos and about 500, under Gen. Castillo, at the little town of Cujababo, a few miles east of Baiquiri. I accepted his offer, impressing it upon him that I could exercise no military control over him except, such as he would concede, and as long as he served under me I would furnish him rations and ammunition.

DISEMBARKATION IN CUBA

Ever since the receipt of my orders I had made a study of the terrain surrounding Santiago, gathering information mainly from the former residents of the city, several of whom were on the transports with me. At this interview all the possible points of attack were for the last time carefully weighed, and then, for the information and guidance of Admiral Sampson and Gen. Garcia, I outlined the plan of campaign, which was as follows:

With the assistance of the small boats of the Navy, the dis-embarkation was to commence on the morning of the 22nd at Baiquiri; on the 21st 500 insurgent troops were to be trans-ferred from Aserraderos to Cujababo, increasing the force al-ready there to 1,000 men. This force, under Gen. Castillo, was to attack the Spanish force at Baiquiri in the rear at the time of disembarkation. This movement was successfully made. To mislead the enemy as to the real point of our intended landing, I requested Gen. Garcia to send a small force (about 500 men), under Gen. Rabi, to attack the little town of Cabanas, situated on the coast a few miles to the west of the entrance to Santiago harbour, and where it was reported the enemy had several men entrenched, and from which a trail leads around the west side of the bay to Santiago.

I also requested Admiral Sampson to send several of his war-ships, with a number of my transports, opposite this town, for the purpose of making a show of disembarking there.

In addition, I asked the admiral to cause a bombardment to be made at Cabanas and also at the forts around the Morro and at the towns of Aguadores, Siboney, and Baiquiri. The troops under Gen. Garcia remaining at Aserraderos were to be trans-ferred to Baiquiri or Siboney on the 24th. This was successfully accomplished at Siboney.

These movements committed me to approaching Santiago from the east over a narrow road, at first in some places not better than a trail, running from Baiquiri through Siboney and Sevilla, and making attack from that quarter. This, in my judg-ment, was the only feasible plan, and subsequent information and results confirmed my judgment.

On the morning of the 22nd the Army commenced to dis-embark at Baiquiri. The following extract from the general or-der indicates the manner in which the troops left the transports and the amount of supplies carried immediately with them:

Headquarters 5th Army Corps
On board S. S. *Seguransa*, At Sea
June 20, 1898

G. O. 18.

1. Under instructions to be communicated to the proper commanders, troops will disembark in the following order:
First—The 2nd Division, 5th Corps (Lawton's). The Gatling Gun Detachment will accompany this division.
Second—Gen. Bates' Brigade. This brigade will form as a reserve to the 2nd Division, 5th Corps.
Third—The dismounted cavalry division (Wheeler's).
Fourth—The 1st Division, 5th Corps (Kent's).
Fifth—The squadron of the 2nd Cavalry (Rafferty's).
Sixth—If the enemy in force vigorously resist the landing, the light artillery, or a part of it, will be disembarked by the battalion commander, and brought to the assistance of the troops engaged. If no serious opposition be offered this artillery will be unloaded after the mounted squadron.
2. All troops will carry on the person the blanket-roll (with shelter-tent and poncho), three days' field rations (with coffee, ground), canteens filled, and 100 rounds of ammunition per man. Additional ammunition, already issued to the troops, tentage, baggage, and company cooking utensils will be left under charge of the regimental quartermaster, with one non-commissioned officer and two privates from each company.
3. All persons not immediately on duty with and constituting a part of the organizations mentioned in the foregoing paragraphs will remain aboard ship until the landing be accomplished, and until notified they can land.
4. The chief quartermaster of the expedition will control all small boats and will distribute them to the best advantage to disembark the troops in the order indicated in paragraph 1.
5. The ordnance officer—2nd Lieut. Brooke, 4th Infantry—

will put on shore at once 100 rounds of ammunition per man, and have it ready for distribution on the firing-line.

6. The commanding general wishes to impress officers and men with the crushing effect a well-directed fire will have upon the Spanish troops. All officers concerned will rigidly enforce fire discipline, and will caution their men to fire only when they can be see the enemy.

By command of Maj.-Gen. Shafter.

E. J. McClernand

A. A. G.

The small boats belonging to the Navy and the transports, together with a number of steam launches, furnished by the Navy, were brought alongside and loaded with troops as prescribed in the order just quoted. When Gen. Lawton's Division was fairly loaded in the small boats, the latter were towed in long lines by the steam launches toward the shore. The sea was somewhat rough, but by the exercise of caution and good judgment the beach was reached and the troops disembarked satisfactorily. As a precaution against a possible attack upon the part of any Spaniards who might have been hidden in the adjacent block-houses and woods, the Navy opened a furious cannonade on these places while the troops were moving toward the shore. It was learned afterward that the Spanish garrison had retired in the direction of Siboney soon after daylight.

By night about 6,000 troops were on shore. Gen. Lawton was ordered to push down a strong force to seize and hold Siboney.

On the 23rd the disembarkation was continued and about 6,000 more men landed. Early on this date Gen. Lawton's advance reached Siboney, the Spanish garrison of about 600 men retiring as he came up, and offering no opposition except a few scattering shots at long range. Some of the Cuban troops pursued the retreating Spaniards and skirmished with them. During the afternoon of this date the disembarkation of Kent's Division was commenced at Siboney, which enabled me to establish a base eight miles nearer Santiago and to continue the unloading of troops and supplies at both points.

The disembarkation was continued throughout the night of the 23rd and 24th, and by the evening of the 24th the disembarkation of my command was practically completed.

PREPARING FOR THE ADVANCE

The orders for June 24th contemplated Gen. Lawton's Division taking a strong defensive position a short distance from Siboney, on the road to Santiago; Kent's Division was to be held near Santiago, where he disembarked; Bates' Brigade was to take position in support of Lawton, while Wheeler's Division was to be somewhat to the rear on the road from Siboney to Baiquiri. It was intended to maintain this situation until the troops and transportation were disembarked and a reasonable quantity of necessary supplies landed. Gen. Young's Brigade, however, passed beyond Lawton on the night of the 232nd-24th, thus taking the advance, and on the morning of the latter date became engaged with a Spanish force entrenched in a strong position at La Guasima, a point on the Santiago road about three miles from Siboney. Gen. Young's force consisted of one squadron of the 1st Cavalry, one of the 10th Cavalry, and two of the 1st United States Volunteer Cavalry; in all, 964 officers and men.

The enemy made an obstinate resistance, but were driven from the field with considerable loss. Our own loss was 1 officer and 15 men killed, 6 officers and 46 men wounded. The reported losses of the Spaniards were 9 killed and 27 wounded. The engagement had an inspiring effect upon our men and doubtless correspondingly depressed the enemy, as it was now plainly demonstrated to them that they had a foe to meet who would advance upon them under a heavy fire delivered from entrenchments. Gen. Wheeler, division commander, was present during the engagement and reports that our troops, officers and men, fought with the greatest gallantry. His report is attached, marked "A". This engagement gave us a well-watered country farther to the front on which to encamp our troops.

My efforts to unload transportation and subsistence stores, so that we might have several days' rations on shore, were continued during the remainder of the month. In this work I was ably seconded by Lieut.-Col. Charles F. Humphrey, deputy Q. M. G., U. S. A., chief quartermaster, and Col. John F. Weston, A. O. G. S., chief commissary; hut, notwithstanding the utmost efforts, it was difficult to land supplies in excess of those required daily to feed the men and animals, and the loss of the scow, mentioned as having broken away during the voyage, as well as the loss at sea of lighters sent by Quartermaster's Department was greatly felt. Indeed, the lack of steam launches, lighters, scows, and wharves can only be appreciated by those who were on the ground directing the disembarkation and landing of supplies. It was not until nearly two weeks after the army landed that it was possible to place on shore three days' supplies In excess of those required for the daily consumption.

After the engagement at La Guasima, and before the end of the month, the army, including Gen. Garcia's command, which had been brought on transports to Siboney from Aserraderos, was mostly concentrated at Sevilla, with the exception of the necessary detachments at Baiquiri and Siboney.

On June 30th I reconnoitred the country about Santiago and made my plan of attack. From a high hill, from which the city was in plain view, I could see the San Juan Hill and the country about El Caney. The roads were very poor, and, indeed, little better than bridle-paths until the San Juan River and El Caney were reached.

The position of El Caney, to the northeast of Santiago, was of great importance to the enemy as holding the Guantanamo road, as well as furnishing shelter for a strong outpost that might be used to assail the right flank of any force operating against San Juan Hill.

In view of this, I decided to begin the attack next day at El Caney with one division, while sending two divisions on the direct road to Santiago, passing by the El Pozo house, and as a diversion to direct a small force against Aguadores, from Siboney along the railroad by the sea, with a view of attracting the attention of the Spaniards in the latter direction and of preventing them from attacking our left flank.

During the afternoon I assembled the division commanders and explained to them my general plan of battle. Lawton's Division, assisted by Capron's Light Battery, was ordered to move out during the afternoon toward El Caney, to begin the attack there early the next morning. After carrying El Caney, Lawton was to move by the El Caney road toward Santiago, and take position on the right of the line. Wheeler's Division of dismounted cavalry, and Kent's Division of infantry, were directed on the Santiago road, the head of the column resting near El Pozo, toward which heights Grimes' Battery moved on the afternoon of the 30th, with orders to take position thereon early the next morning, and at the proper time prepare the way for the advance of Wheeler and Kent, on San Juan Hill. The attack at this point was to be delayed until Lawton's guns were heard at El Caney and his infantry fire showed he had become well engaged.

The remainder of the afternoon and night was devoted to cutting out and repairing the roads, and other necessary preparations for battle. These preparations were far from what I desired them to be, but we were in a sickly climate; our supplies had to be brought forward by a narrow wagon road, which the rains might at any time render impassable; fear was entertained that a storm might drive the vessels containing our stores to sea, thus separating us from our base of supplies; and, lastly, it was reported that Gen. Pando, with 8,000 reinforcements for the enemy, was *en route* from Manzanillo, and might be expected in a few days. Under these conditions, I determined to give battle without delay.

The Battle of El Caney

Early on the morning of July 1st, Lawton was in position around El Caney, Chaffee's Brigade on the right, across the Guantanamo road, Miles' Brigade in the centre, and Ludlow's on the left. The duty of cutting off the enemy's retreat along the Santiago road was assigned to the latter brigade. The artillery opened on the town at 6:15 a. m. The battle here soon became general, and was hotly contested. The enemy's position was naturally strong, and was rendered more so by blockhouses, a stone fort, and entrenchments cut in solid rock, and the loop-holing of a solidly built stone church. The opposition offered by the enemy was greater than had been anticipated, and prevented Lawton from joining the right of the main line during the day, as had been intended. After the battle had continued for some time, Bates' Brigade of two regiments reached my headquarters from Siboney. I directed him to move near El Caney, to give assistance if necessary. He did so, and was put in position between Miles and Chaffee. The battle continued with varying intensity during most of the day and until the place was carried by assault about 4:30 p. m. As the Spaniards endeavoured to retreat along the Santiago road, Ludlow's position enabled him to do very effective work, and to practically cut off all retreat in that direction.

After the battle at El Caney was well opened, and the sound of the small-arm fire caused us to believe that Lawton was driving the enemy before him, I directed Grimes' Battery to open fire from the heights of El Pozo on the San Juan block-house, which could be seen situated in the enemy's entrenchments extending along the crest of San Juan Hill. This fire was effective, and the enemy could be seen running away from the vicinity of the block-house. The artillery fire from El Pozo was soon returned by the enemy's artillery. They evidently had the range of this hill, and their first shells killed and wounded several men. As the Spaniards used smokeless powder, it was very difficult to locate the position of their pieces, while, on the contrary, the smoke caused by our black powder plainly indicated the position of our battery.

At this time the cavalry division, under Gen. Sumner, which was lying concealed in the general vicinity of the El Pozo house, was ordered forward with directions to cross the San Juan River and deploy to the right of the Santiago side, while Kent's Division was to follow closely in its rear and deploy to the left.

These troops moved forward in compliance with orders, but the road was so narrow as to render it impracticable to retain the column of fours formation at all points, while the undergrowth on either side was so dense as to preclude the possibility of deploying skirmishers. It naturally resulted that the progress made was slow, and the long-range rifles of the enemy's infantry killed and wounded a number of our men while marching along this road, and before there was any opportunity to return this fire. At this time Generals Kent and Sumner were ordered to push forward with all possible haste and place their troops in position to engage the enemy. Gen. Kent, with this end in view, forced the head of his column alongside of the cavalry column as far as the narrow trail permitted, and thus hurried his arrival at the San Juan and the formation beyond that stream. A few hundred yards before reaching the San Juan the road forks, a fact that was discovered by Lieut.-Col. Derby of my staff, who had approached well to the front in a war balloon. This information he furnished to the troops, resulting in Sumner moving on the right-hand road, while Kent was enabled to utilize the road to the left.

Gen. Wheeler, the permanent commander of the cavalry division, who had been ill, came forward during the morning, and later returned to duty and rendered most gallant and efficient service during the remainder of the day.

After crossing the stream, the cavalry moved to the right with a view of connecting with Lawton's left, when he could come up, and with their left resting near the Santiago road.

In the meantime Kent's Division, with the exception of two regiments of Hawkins' Brigade, being thus uncovered, moved rapidly to the front from the forks previously mentioned in the road, utilizing both trails, but more especially the one to the left, and, crossing the creek, formed for attack in front of San Juan Hill. During the formation the 2nd Brigade suffered severely. While personally superintending this movement, its gallant commander, Col. Wikoff, was killed. The command of the brigade then devolved upon Lieut.-Col. Worth, 13th Infantry, who was soon severely wounded, and next upon Lieut.-Col. Liscum, 24th Infantry, who, five minutes later, also fell under the terrible fire of the enemy, and the command of the brigade then devolved upon Lieut.-Col. Ewers, 9th Infantry.

While the formation just described was taking place, Gen. Kent took measures to hurry forward his rear brigade. The 10th and 2nd Infantry were ordered to follow. Wikoff's Brigade, while the 21st was sent on the right-hand road to support the 1st Brigade, under Gen. Hawkins, who had crossed the stream and formed on the right of the division. The 2nd and 10th Infantry, Col. E. P. Pearson commanding, moved forward in good order on the left of the division, passed over a green knoll, and drove the enemy back toward his trenches.

After completing their formation under a destructive fire, and advancing a short distance, both divisions found in their front a wide bottom, in which had been placed a barbed-wire entanglement, and beyond which there was a high hill, along the crest of which the enemy was strongly posted. Nothing daunted, these gallant men pushed on to drive the enemy from his chosen position, both divisions losing heavily. In this assault Col. Hamilton, Lieuts. Smith and Shipp were killed, and Col. Carroll, Lieuts. Thayer and Myer, all in the cavalry, were wounded.

Great credit is due to Brig.-Gen. H. S. Hawkins, who, placing himself between his regiments, urged them on by voice and bugle calls to the attack so brilliantly executed.

In this fierce encounter words fail to do justice to the gallant regimental commanders and their heroic men, for, while the generals indicated the formations and the points of attack, it was, after all, the intrepid bravery of the subordinate officers and men that planted our colours on the crest of San Juan Hill and drove the enemy from his trenches and block-houses, thus gaining a position which sealed the fate of Santiago.

In this action on this part of the field most efficient service was rendered by Lieut. John H. Parker, 13th Infantry, and the Gatling Gun Detachment under his command. The fighting continued at intervals until nightfall, but our men held resolutely to the positions gained at the cost of so much blood and toil.

I am greatly indebted to Gen. Wheeler, who, as previously stated, returned from the sick-list to duty during the afternoon. His cheerfulness and aggressiveness made itself felt on this part of the battle-field, and the information he furnished to me at various stages of the battle proved to be most useful.

My own health was impaired by overexertion in the sun and intense heat of the day before, which prevented me from participating as actively in the battle as I desired; but from a high hill near my headquarters I had a general view of the battle-field, extending from El Caney on the right to the left of our lines on San Juan Hill. His staff officers were stationed at various points on the field, rendering frequent reports, and through them, by the means of orderlies and the telephone, I was enabled to transmit my orders. During the afternoon I visited the position of Grimes' Battery on the heights of El Pozo, and saw Sumner and Kent in firm possession of San Juan Hill, which I directed should be entrenched during the night. My engineer officer, Lieut.-Col. Derby, collected and sent forward the necessary tools, and during the night trenches of very considerable strength were constructed. During the afternoon, Maj. Dillenback, by my order, brought forward the two remaining batteries of his battalion and put them in position at El Pozo, to the left of Grimes. Later in the afternoon all three batteries were moved forward to positions near the firing-line, but the nature of the country and the intensity of the enemy's small-arm fire was such that no substantial results were gained by our artillery in the new positions. The batteries were entrenched during the night. Gen. Duffield, with the 33rd Michigan, attacked Aguadores, as ordered, but was unable to accomplish more than to detain the Spaniards in that vicinity.

After the brilliant and important victory gained at El Caney, Lawton started his tried troops, who had been fighting all day and marching much of the night before, to connect with the right of the cavalry division. Night came on before this movement could be accomplished. In the darkness the enemy's pickets were encountered, and the division commander, being uncertain of the ground and as to what might be in his front, halted his command and reported the situation to me. This information was received about 12:30 a. m., and I directed Gen. Lawton to return by my headquarters and the El Pozo house as the only certain way of gaining his new position.

This was done, and the division took position on the right of the cavalry early next morning; Chaffee's Brigade arriving first, about half-past seven, and the other brigades before noon.

On the night of July 1st, I ordered Gen. Duffield, at Siboney, to send forward the 34th Michigan and the 9th Massachusetts. Both of which had just arrived from the United States. These regiments reached the front the next morning. The 34th was placed in rear of Kent, and the 9th was assigned to Bates, who placed it on his left.

Soon after daylight on July 2nd the enemy opened battle, but because of the entrenchments made during the night, the approach of Lawton's Division, and the presence of Bates' Brigade, which had taken position during the night on Kent's left, little apprehension was felt as to our ability to repel the Spaniards.

It is proper here to state that Gen. Bates and his brigade had performed most arduous and efficient service, having marched much of the night of June 30th-July 1st, and a good part of the latter day, during which he also participated in the battle of El Caney, after which he proceeded, by way of El Pozo, to the left of the line at San Juan, reaching his new position about midnight.

All day on the 2nd the battle raged with more or less fury, but such of our troops as were in position at daylight held their ground, and Lawton gained a strong and commanding position on the right. About 10 p. m. the enemy made a vigorous assault to break through my lines, but he was repulsed at all points.

Summoning the Enemy to Surrender

On the morning of the 3rd the battle was renewed, but the enemy seemed to have expended his energy in the assault of the previous night, and the firing along the lines was desultory until stopped by my sending the following letter within the Spanish lines:

Headquarters U. S. Forces
Near San Juan River
July 3, 1898—8:30 a. m.
Sir,
I shall be obliged, unless you surrender, to shell Santiago

189

de Cuba. Please inform the citizens of foreign countries, and all the women and children, that they should leave the city before 10 o'clock tomorrow morning.

Very respectfully, your obedient servant,

William R. Shafter

Maj.-Gen. U. S. Vols.

The Commanding General of the Spanish Forces, Santiago de Cuba

To this letter I received the following reply:

Santiago de Cuba

July 3, 1898

His Excellency the General Commanding Forces of the United States

Near San Juan River

Sir,

I have the honour to reply to your communication of to-day, written at 8:30 a. m. and received at 1 p. m., demanding the surrender of this city, or, in the contrary case, announcing to me that you will bombard this city, and that I advise the foreigners, women and children, that they must leave the city before 10 o'clock tomorrow morning.

It is my duty to say to you that this city will not surrender, and that I will inform the foreign consuls and inhabitants of the contents of your message.

Very respectfully

Jose Toral

Commander-in-Chief 4th Corps

Several of the foreign consuls came into my lines and asked that the time given for them—the women and children—to depart from the city be extended until 10 o'clock on July 5th. This induced me to write a second letter, as follows:

Santiago de Cuba

July 3, 1898

Sir,—In consideration of a request of the consular officers in your city for further delay in carrying out my intentions to fire on the city, and in the interests of the poor women and children who will suffer very greatly by their hasty and enforced departure from the city, I have the honour to announce that I will delay such action, solely in their interests, until noon of the 5th, provided that during the interim your forces make no demonstration whatever upon those of my own.

I am, with great respect, your obedient servant,
William R Shafter
Maj.-Gen. U. S. A.

The Commanding General, Spanish Forces

My first message went under a flag of truce at 12:42 p.m. I was of the opinion that the Spaniards would surrender if given a little time, and I thought this result would be hastened if the men of their army could be made to understand they would be well treated as prisoners of war. Acting upon this presumption, I determined to offer to return all the wounded Spanish officers at El Caney who were able to bear transportation, and who were willing to give their paroles not to serve against the forces of the United States until regularly exchanged. This offer was made and accepted. These officers, as well as several of the wounded Spanish privates, twenty-seven in all, were sent to their lines under the escort of some of our mounted cavalry. Our troops were received with honours, and I have every reason to believe the return of the Spanish prisoners produced a good impression on their comrades.

Operations After Santiago—Our Losses

The cessation of firing about noon on the 3rd practically terminated the battle of Santiago; all that occurred after this time may properly be treated under the head of the siege which followed. After deducting the detachments required at Siboney and Baiquiri to render those depots secure from attack, organizations held to protect our flanks, others acting as escorts and guards to light batteries, the members of the Hospital Corps, guards left in charge of blanket-rolls which the intense heat caused the men to cast aside before entering battle, orderlies, etc., it is doubtful if we had more than 12,000 men on the firing-line on July 1, when the battle was fiercest and when the important and strong positions of El Caney and San Juan were captured.

A few Cubans assisted in the attack at El Caney, and fought valiantly, but their numbers were too small to materially change the strength, as indicated above. The enemy confronted us with numbers about equal to our own; they fought obstinately in strong and entrenched positions, and the results obtained clearly indicate the intrepid gallantry of the company officers and men, and the benefits derived from the careful training and instruction given in the company in the recent years in rifle practice and other battle exercises. Our losses in these battles were 22 officers and 208 men killed, and 81 officers and 1,203 men wounded; missing, 79. The missing, with few exceptions, reported later.

The arrival of Gen. Escario on the night of July 2nd, and his entrance into the city was not anticipated, for although it was known, as previously stated, that Gen. Pando had left Manzanillo with reinforcements for the garrison of Santiago, it was not believed his troops could arrive so soon. Gen. Garcia, with between 4,000 and 5,000 Cubans, was entrusted with the duty of watching for and intercepting the reinforcement expected. This, however, he failed to do, and Escario passed into the city along on my extreme right and near the bay. Up to this time I had been unable to complete investment of the town with my own men; but to prevent any more reinforcements coming in or the enemy escaping. I extended my lines as rapidly as possible to the extreme right, and completed the investment of the place, leaving Gen. Garcia's forces in the rear of my right flank to scout the country for any approaching Spanish reinforcements, a duty which his forces were very competent to perform.

It had been reported that 8,000 Spanish troops had left Holquin for Santiago. It was also known that there was a considerable force at San Luis, twenty miles to the north.

In the battle of Santiago the Spanish navy endeavoured to shell our troops on the extreme right, but the latter were concealed by the inequalities of the ground, and the shells did little, if any, harm. Their naval forces also assisted in the trenches, having 1,000 on shore, and I am informed they sustained considerable loss; among others, Admiral Cervera's chief-of-staff was killed. Being convinced that the city would fall, Admiral Cervera determined to put to sea, informing the French consul it was better to die fighting than to sink his ships. The news of the great naval victory which followed was enthusiastically received by the Army.

The information of our naval victory was transmitted under flag of truce to the Spanish commander in Santiago on July 4th, and the suggestion again made that he surrender to save needless effusion of blood.

On the same date I informed Admiral Sampson that if he would force his way into the harbour the city would surrender without any further sacrifice of life. Commodore Watson replied that Admiral Sampson was temporarily absent, but that in his (Watson's) opinion the Navy should not enter the harbour.

In the meanwhile letters passing between Gen. Toral and myself caused the cessation of hostilities to continue. Each army, however, continued to strengthen its entrenchments. I was still of the opinion the Spaniards would surrender without much more fighting, and on July 6th called Gen. Toral's attention to the changed conditions, and at his request gave him time to consult his home government. This he did, asking that the British consul, with the employees of the cable company, be permitted to return from El Caney to the city. This I granted.

The strength of the enemy's position was such I did not wish to assault if it could be avoided.

An examination of the enemy's works, made after the surrender, fully justifies the wisdom of the course adopted. The entrenchments could only have been carried with very great loss of life, probably with not less than 6,000 killed and wounded.

Negotiations with General Toral

On July 8th Gen. Toral offered to march out of the city with arms and baggage, provided he would not be molested before reaching Holquin, and to surrender to the American forces the territory then occupied by him. I replied that while I would submit his proposition to my home government. I did not think it would be accepted.

In the meanwhile arrangements were made with Admiral Sampson that when the Army again engaged the enemy the Navy would assist by shelling the city from ships stationed off Aguadores, dropping a shell every few minutes.

On July 10th the 1st Illinois and the 1st District of Columbia arrived and were placed on the line to the right of the Cavalry division. This enabled me to push Lawton farther to the right and to practically command the Cobre road.

On the afternoon of the date last mentioned the truce was broken off at 4 p.m., and I determined to open with four batteries of artillery and went forward in person to the trenches to give the necessary orders, but the enemy anticipated us by opening fire with his artillery a few minutes after the hour stated. His batteries were apparently silenced before night, while ours continued playing upon his trenches until dark. During this firing the Navy fired from Aguadores, most of the shells falling in the city. There was also some small arms firing. On this afternoon and the next morning, we lost Capt. Charles W. Rowell, 2nd Infantry, and one man killed, and Lieut. Lutz, 2nd Infantry, and ten men wounded.

On the morning of July 11th the bombardment by the Navy and my field guns was renewed, and continued until nearly noon, and on the same day I reported to the Adjutant General of the Army that the right of Ludlow's brigade of Lawton's division rested on the bay. Thus our hold upon the enemy was complete.

At 2 p. m. on this date, the 11th, the surrender of the city was again demanded. The firing ceased, and was not again renewed. By this date the sickness in the Army was increasing very rapidly, as a result of exposure in the trenches to the intense heat of the sun and the heavy rains. Moreover, the dews in Cuba are almost equal to rains. The weakness of the troops was becoming so apparent I was anxious to bring the siege to an end, but in common with most of the officers of the Army,

I did not think an assault would be justifiable, especially as the enemy seemed to be acting in good faith in their preliminary propositions to surrender.

On July 11th I wrote to General Toral as follows:

> With the largely increased forces which have come to me and the fact that I have your line of retreat securely in my hands, the time seems fitting that I should again demand of your excellency the surrender of Santiago and of your excellency's army. I am authorized to state that should your excellency so desire, the Government of the United States will transport the entire command of your excellency to Spain.

General Toral replied that he had communicated my proposition to his General-in-Chief, General Blanco.

July 12th I informed the Spanish commander that Major General Miles, Commander-in-Chief of the American Army, had just arrived in my camp, and requested him to grant us a personal interview on the following day. He replied he would be pleased to meet us. The interview took place on the 13th, and I informed him his surrender only could be considered, and that as he was without hope of escape he had no right to continue the fight.

On the 14th another interview took place, during which General Toral agreed to surrender, upon the basis of his army, the 4th Army Corps, being returned to Spain, the capitulation embracing all of Eastern Cuba, east of a line passing from Aserraderos, on the south, to Sagua de Tanamo, on the north, via Palma, Soriano. It was agreed Commissioners should meet during the afternoon to definitely arrange the terms of surrender, and I appointed Major Generals Wheeler and Lawton and Lieutenant Miley to represent the United States.

The Spanish Commissioners raised many points, and were especially desirous of retaining their arms. The discussion lasted until late at night and was renewed at 9:30 o'clock next morning. The terms of surrender finally agreed upon included about 12,000 Spanish troops in the city and as many more in the surrendered district.

It was arranged that the formal surrender should take place between the lines on the morning of July 17th, each army being represented by 100 armed men. At the time appointed, I appeared at the place agreed upon, with my general officers, staff, and 100 troopers of the 2nd Cavalry, under Captain Brett. General Toral also arrived with a number of his officers and 100 infantry. We met midway between the representatives of our two Armies, and the Spanish commander formally consummated the surrender of the city and the 24,000 troops in Santiago and the surrendered district. After this ceremony I entered the city with my staff and escort, and at 12 o'clock, noon, the American flag was raised over the Governor's palace with appropriate ceremonies.

The 9th Infantry immediately took possession of the city and perfect order was maintained. The surrender included a small gunboat and about 200 seamen, together with five merchant ships in the harbour. One of these vessels, the Mexico, had been used as a war vessel, and had four guns mounted on it.

In taking charge of the civil government, all officials who were willing to serve were retained in office, and the established order of government was preserved as far as consistent with the necessities of military rule.

I soon found the number of officials was excessive, and I greatly reduced the list, and some departments were entirely abolished.

A collector of customs, Mr. Donaldson, arrived soon after the surrender, and, due to his energy and efficiency, this department was soon working satisfactorily. The total receipts had, up to my departure, been $102,000.

On August 4th I received orders to begin the embarkation of my command and ship them to Montauk Point Long Island, New York. The movement continued without interruption until August 25th, when I sailed for Montauk with the last troops in my command, turning over the command of the district to Major General Lawton.

Difficulties Encountered in the Campaign

Before closing my report I wish to dwell upon the natural obstacles I had to encounter and which no foresight could have overcome or obviated. The rocky and precipitous coast afforded no sheltered landing places, the roads were mere bridle-paths, the effect of the tropical sun and rains upon the unacclimated troops was deadly, and a dread of strange and unknown diseases had its effect on the Army.

At Baiquiri the landing of the troops and stores was made a small wooden wharf, which the Spaniards tried to burn, but unsuccessfully, and the animals were pushed into the water and guided to a sandy beach about 200 yards in extent. At Siboney the landing was made on the beach and at a small wharf erected by the engineers.

I had neither the time nor the men to spare to construct permanent wharves.

In spite of the fact that I had nearly 1,000 men continuously at work on the roads, they were at times impassable for wagons.

The San Juan and Aguadores rivers would often suddenly rise so as to prevent the passage of wagons, and then the eight pack trains with the command had to be depended upon for the victualing of my Army, as well as the 20,000 refugees, who could not in the interests of humanity be left to starve while we had rations.

Often for days nothing could be moved except on pack trains.

After the great physical strain and exposure of July 1st and 2nd, the malarial and other fevers began to rapidly advance throughout the command, and on July 4th the yellow fever appeared at Siboney. Though efforts were made to keep this fact from the Army, it soon became known.

The supply of Quartermaster and Commissary stores during the campaign was abundant, and notwithstanding the difficulties in landing and transporting the ration, the troops on the firing line were at all times supplied with its coarser components, namely, of bread, meat, sugar, and coffee.

There was no lack of transportation, for at no time up to the surrender could all the wagons I had be used.

In reference to the sick and wounded, I have to say that they received every attention that was possible to give them. The medical officers, without exception, worked night and day to alleviate the suffering, which was no greater than invariably accompanies a campaign. It would have been better if we had had more ambulances, but as many were taken as was thought necessary, judging from previous campaigns.

The discipline of the command was superb, and I wish to invite attention to the fact that not an officer was brought to trial by court martial, and, as far as I know, no enlisted men. This speaks volumes for an Army of this size and in a campaign of such duration.

In conclusion, I desire to express to the members of my staff my thanks for their efficient performance of all the duties required of them, and the good judgment and bravery displayed on all occasions when demanded.

I submit the following recommendations for promotion, which I earnestly desire to see made. It is a very little reward to give them for their devotion and fearless exposure of their lives in their country's cause:

E. J. McClernand, Lieutenant Colonel and Adjutant General, U. S. A., to be brevetted Colonel for gallantry in the face of the enemy on the 1st and 2nd of July, and to be brevetted Brigadier General for faithful and meritorious service throughout the campaign.

Geo. McC. Derby, Lieutenant Colonel of Engineers, U. S. V., to be brevetted Colonel for hazardous service on July 1st and 2nd in reconnoitring the enemy's lines, and to be brevetted Brigadier General for hazardous and meritorious service in ascending, under a hot fire, in a war balloon on July 1st, thus gaining valuable information.

J. D. Miley, Lieutenant Colonel and Inspector General, U. S. A., to be brevetted Colonel for conspicuous gallantry in the battle of San Juan on July 1st, and to be brevetted Brigadier General for faithful and meritorious service throughout the campaign.

R. H. Noble, Major and Adjutant General, U. S. V., to be brevetted Lieutenant Colonel for faithful and meritorious service throughout the campaign.

J. J. Astor, Lieutenant Colonel and Inspector General, U. S. V., to be brevetted Colonel for faithful and meritorious service during the campaign.

B. F. Pope, Lieutenant Colonel and Surgeon, U. S. V., to be brevetted Colonel for faithful and meritorious service during the campaign.

Maj. S. W. Groesbeck, Judge Advocate, U. S. A., to be brevetted Lieutenant Colonel for faithful and meritorious service throughout the campaign.

Charles F. Humphrey, Lieutenant Colonel, Quartermaster's Department, to be brevetted Brigadier General for faithful and meritorious service throughout the campaign.

John F. Weston, Colonel and Assistant Commissary General of Subsistence, Chief Commissary, to be brevetted Brigadier General for meritorious service throughout the campaign.

C. G. Starr, Major and Inspector General, U. S. V., to be brevetted Lieutenant Colonel for faithful and meritorious service throughout the campaign.

Leon Roudiez, Major and Quartermaster, U. S. V., to be brevetted Lieutenant Colonel for faithful and meritorious conduct throughout the campaign.

H. J. Gallagher, Major and Commissary of Subsistence, U. S. V., to be brevetted Lieutenant Colonel for faithful and meritorious service throughout the campaign.

Capt. Brice, Commissary of Subsistence, U. S. V., to be brevetted Major for faithful and meritorious service throughout the campaign.

E. H. Plummer, Captain, U. S. A., A. D. C., to be brevetted Major for faithful and meritorious service throughout the campaign.

J. C. Gilmore, Jr., Captain and Assistant Adjutant General, U. S. V., to be brevetted Major for faithful and meritorious service during the campaign.

W. H. McKittrick, Captain and Assistant Adjutant General,

U. S.V., to be brevetted Major for faithful and meritorious service during the campaign.

Capt. Johnson, Assistant Quartermaster, U. S.V., to be brevetted Major for faithful and meritorious service during the campaign.

I wish to invite special attention to Dr. G. Goodfellow, of New York, who accompanied me throughout the campaign and performed much professional service as well as duties as Volunteer aid. I recommend him for favourable consideration of the War Department.

Mr. G. F. Hawkins, of New York, also accompanied me as Volunteer aid, and I recommend him for favourable consideration of the War Department for faithful and important services rendered.

My thanks are due to Admiral Sampson and Captain Goodrich, U. S. N., for their efficient aid in disembarking my Army. Without their assistance it would have been impossible to have landed in the time I did.

I also express my warmest thanks to division, brigade, and regimental commanders, without exception, for their earnest efforts in carrying out my wishes and for the good judgment they invariably displayed in handling their troops.

The reports of the division commanders are attached hereto, and those of the brigade and regimental commanders forwarded herewith, and attention respectfully invited to them. Very respectfully,

Wm. R. Shafter
Major-General, United States Volunteers
Commanding United States Forces in Cuba
Adjutant General of the Army, Washington, D. C.

Appendix 3

Bivouac
near Santiago, Cuba
July 23, 1898
The Adjutant-General
U. S. Army, Washington, D. C.
Sir,

In compliance with orders I have the honour to submit the following report of my command, the Gatling Gun Detachment, 5th Army Corps, covering its operations down to the present date:

1. Organization—Pursuant to instructions from Gen. Shafter I was given a detail of two sergeants and ten men on the 26th of May, 1898, from the 13th Infantry, then in camp near Tampa, Fla., and directed to report to 1st Lieut. John T. Thompson, O. D., ordinance officer, Tampa, "for duty with Gatling guns." I was placed in charge of four guns, model 1895, cal. 30, and at once began the instruction of the detachment. On June 1st I received verbal instruction to assist Lieut. Thompson in his work at the ordinance depot, and performed this duty in addition to my duties with the guns until June 6, 1898, superintending issues to the expedition (5th Corps) then fitting out for Cuba.

On June 6th I took my men and guns aboard the transport *Cherokee*, and on June 11th, per special orders No. 16 of that date, my detail was increased to thirty-seven men, all told, of whom one was left sick in hospital at Tampa. About twelve of these did not join me, however, until after debarkation at Baiquiri, Cuba.

On June 25th I received verbal instructions from Gen. Shafter to disembark at once, select the necessary number of mules (two per gun), and get to the front as soon as possible, reporting on my arrival there to Gen. Wheeler, then in command of all the troops at the front. I was unable to obtain any tentage for myself, and had only shelter-tents for my men.

I was joined on June 25th by Capt. Henry Marcotte, 17th Infantry, retired, regularly authorized correspondent of the Army and Navy Journal, who has been with me ever since, enduring all the vicissitudes of the season with Spartan fortitude, although equally destitute of cover as myself and 60 years of age.

I desire to express here officially and fully, my sincere gratitude for the kindness which permitted him to accompany my command, and the great appreciation of the valuable advice and assistance which he has given continually. His large experience of war, his clear head and good judgment have always been at hand to aid, and his cool example to myself and my men under fire did much to steady us and keep us up to our work when we were first called on to face that ordeal.

All of the detachments, who had not previously joined me, did so on June 26th, on which day I reached the extreme front and reported to Gen. Wheeler. The guns were posted in a position to sweep the neighbouring hills toward the enemy, and I went into camp, remaining there until the morning of July 1st.

Summing up the organization, it should be stated here that the detachment was organized at the first, and has ever since remained an independent command, receiving its orders directly from the corps commander. It has had its own records, returns, rolls, etc., and has been rationed separately all the time, and is composed of men selected by myself from various regiments.

2. The Battery in Action—On the morning of July 1st, I broke camp at 4:30 a.m., and pursuant to instructions from Gen. Shafter, proceeded to El Poso, placing my battery, as I shall henceforth call it, in support behind the position taken by a battery of artillery. I took this position about 6 a.m., and soon after the artillery arrived, went on to battery and opened fire at Santiago, the range being 2,600 yards. After some time the enemy replied

with a well-directed fire, the second shell bursting directly over my battery in rear of artillery. Neither my men nor mules showed any signs of disturbance, and we remained in our perilous position nearly twenty minutes, the enemy's shells bursting all around us, until ordered to the rear by the chief-of-staff.

The battery went to the rear under fire quietly until out of range, and remained there until the artillery fire ceased, at about 9 a.m. Private Hoft, Company D, 13th Infantry, a member of the detachment who had been detailed to guard the camp equipage at El Poso, remained at his post during the whole of the artillery fight, and deserves great credit therefore, his battery having been ordered to the rear. At 9 a.m. I returned to El Poso, and there received the following instructions from Col. McClernand, A. A. G., 5th Corps: "Find the 71st N.Y.V. and go on with them, if you can. If this is not practical, find the best position you can, and use your guns to the best advantage."

Pursuant to these instructions, I went forward about a half-mile and found the 71st N.Y.V. halting to learn what their instructions were. I could get no clear idea of what they were going to do, but waited about fifteen minutes in their rear to find out. Meantime troops continually passed us toward the front. Then, about 10:15, firing began in front. I rode forward alone along the rode, which was a narrow defile through the jungle, and found that about a half-mile in front was a creek, upon the crossing of which the enemy's fire seemed concentrated. In front of this crossing seemed to be a level plain of about 400 to 800 yards, beyond which was a semi-circular ridge crowned with Spanish trenches from which the Spanish fire seemed to come. Men were being hit continually at this place (the ford), but it seemed to me to be a good place to work my battery effectively.

I rode back, finding the Seventy-first still lying beside the road without any apparent intention of moving. I determined to leave them and go into action. Taking a gallop I moved the battery forward nearly to the ford (about 150 yards), where I met Col. Derby of Gen. Shafter's staff, who informed me that the troops were not yet sufficiently deployed to take advantage of my fire, and advised me to wait.

The bullets were cutting through all around, and, as we learned afterward, the enemy's sharpshooters were actually in the woods near us, up in tall trees, picking off officers and men. It should be stated here that the sudden increase of the enemy's fire at this time was caused by a wild cheering set up by the 71st N.Y.V., as the battery passed them on its way to the front. The cheering located our position for the enemy and drew his fire. Many a brave soldier who had gone to the front was put forever beyond the possibility of cheering by this outburst of ignorant enthusiasm.

I acted on Col. Derby's advice, and he promised to send me word when the moment for proper action came. This was necessary, as I knew only part of the plan of battle and might have jeopardized other parts of prematurely exposing our strength at this point. The gun crews lay down under their guns and steadily remained at this posts. The fire finally grew so hot that I moved about 100 yards back. This was about 12, noon. At 1 p.m., or about that time, I received a message sent by Col. Derby, I think, as follows: "Gen. Shafter directs that you give one of your guns to Lieut. Miley, take the others forward beyond the ford where the dynamite gun is, and go into action at the best point you can find." I obeyed the order, giving Lieut. Miley Sergeant Weigle's gun and crew and moving the rest forward at a gallop to the point beyond the ford, which I had already selected as a good place. The battery opened with three guns simultaneously at ranges of 600 to 800 yards at 1:15 p. m.

The enemy at first concentrated his fire upon us, but soon weakened and in five minutes was clambering from his trenches and running to the rear. We fired as rapidly as possible upon the groups thus presented until I saw a white handkerchief waved by some one of my own regiment, the 13th Infantry, and at the same moment Capt. Landis, 1st Cavalry, who had voluntarily assisted me throughout, said: "Better stop; our own men are climbing up the ridge." I ordered the fire to cease at 1:23½ p.m., and a moment later saw our own troops occupy the crest of the hill. The firing had been, continued by the battery until our own troops were within 150 yards of the enemy's trench, a fact made possible by the steep slope of the hill upon which the enemy had been.

At the time when my battery went into action I had no support, and the position I took was at least 100 yards in front of any of our troops along this part of the line. About the time I ceased firing Lieut.-Col. Baldwin, 10th Cavalry, put two troops in support of my battery.

I have advanced in a letter to the Adjutant General from Fort Leavenworth, dated January 1st, 1898, the theory that such guns as these can be used offensively. The conditions of this assault were favourable, the morale of my men superb, and the use made of the guns followed the theory therein set forth with the exactness of a mathematical demonstration. The infantry and cavalry had been pounding away for two hours on these positions; in eight and one-half minutes after the Gatlings opened the works were ours. Inspired by the friendly rattle of the machine guns, our own troops rose to the charge; while the enemy amazed by our sudden and tremendous increase of fire, first diverted his fire to my battery, and then, unable to withstand the hail of bullets, augmented by the moral effect of our battery fire and the charging line, broke madly from his safe trenches and was mercilessly cut by fire from these guns during his flight.

I at once limbered up and took stock of my losses. One man was killed, one badly wounded, one mule hit twice, but not much injured, and several men were missing.

Suddenly the fire was resumed at the front. I moved my three pieces forward again at a gallop, and went into action on the skirmish line on top of the captured position, with two pieces to the right and one to the left of the main road from El Poso to Santiago. I was compelled to make the skirmishers give way to the right and left in order to get room for my guns on the firing-line, and to impress stragglers to carry ammunition.

Capt. Ayres, 10th Cavalry, gave me a detail of one sergeant and two privates, all of whom did fine service. It seemed to me that the enemy was trying to retake the position. About 4 to 4:14 p. m. I saw a body, apparently about 400, of the enemy to the right front of my position, apparently in front of the position occupied by Lieut.-Col. Roosevelt with the 1st Volunteer Cavalry. I turned a Gatling gun on them, using 600-yard range, and they disappeared. Soon after the firing sensibly slackened.

In the rapid fire on this last body of the enemy I had over-heated one piece, and it went temporarily out of action. I went over to Col. Roosevelt's position, about a quarter of a mile to the right of a salient, and reconnoitred. While there Sergeant Weigle reported to me with his piece, informing me that Lieut. Miley had not put it into action, and asked for instructions. This was about the hour of 5 p. m., and the fire became warmer at that moment. I directed Sergeant Weigle to run his piece up on the firing-line and to report to the officer in charge thereof. He did so and went into action at once. Col. Roosevelt, who was and remained present, informs me that the gun was very effectively used. I rejoined my other two guns and put both of them on the line at the left of the El Poso road. At sundown the enemy made a sharp attack, and all three of my guns were effectively used.

During the fight a battery in the city opened on my two guns, firing 16 cm. shells. I at once turned my guns on it and kept up so warm a fire that the cannoneers left their battery and did not return. In all they had fired three shells at us, all of which broke just over or beyond the battery. I secured the fuse of one, still warm, and after the surrender visited the battery which had fired at us and examined the gun. It is a 16 cm. (6.2992 inches) bronze rifle gun in barbette on a pintle. This is probably the first time in land fighting that such a piece was ever silenced by machine-gun fire. The range I used was 2,000 yards (estimated).

The guns were used during the remainder of the fighting in the trenches. I took off the wheels and put the guns on the car-riages in emplacements, erecting a sandbag parapet in front as cover during the night of July 4th. The disabled gun was brought up and repaired, subsequently participating in the fighting. The dynamite gun, under Sergeant Borrowe, 1st Volunteer Cavalry, cooperated with the battery thus formed, and the whole battery, including the two Colt automatic rapid-fire guns under Lieut. Tiffany, 1st United States Volunteer Cavalry, did good work in all the subsequent fighting.

I supplied about eight thousand rounds of captured Mauser cartridges to Tiffany, which had been captured by my battery, and which he used effectively in his Colt's guns. I had a strong fire directed upon a battery of seven pieces of the enemy's artillery at a distance of 1,500 yards in front every time any attempt was made to use this battery. The result was that only three shots were fired from these guns after July 4th. I visited this battery after the surrender and found every gun in working order, the 16 cm. gun being actually loaded. As no organization, except my battery, of which I had general direction, had such orders, so far as I can learn, the conclusion is that this battery of machine guns kept out of action seven pieces of the enemy's artillery by making it too warm for his gunners to stay in their batteries.

I have made certain recommendations in hasty reports for gallantry, which I personally witnessed. They were as follows:

Capt. J. R. F. Landis, 1st Cavalry, medal of honour. Volunteered to assist observation of fire July 1st, and rendered great service at imminent peril of his life made necessary in order to render such service.

Sergeant John N. Weigle, 9th Infantry, 2nd Lieutenant U. S. Army (regulars). For conspicuous daring, intelligence, and coolness in action, July 1st.

Corporal Charles C. Steigenwald, 13th Infantry, 2nd Lieutenant U. S. Army (regulars). For coolness and judgment in keeping his gun in action with only one man to help on July 1st.

Private Fred C. Elkins, 17th Infantry, 2nd Lieutenant United States Volunteers. For conspicuous daring and courage in action. Although wounded, he remained at his post until he fell from exhaustion, July 1st.

Corporal Matthew Doyle, 13th Infantry, medal of honour. Conspicuous gallantry and coolness in action. When, two men had been shot down by his side he continued to work his gun effectively alone until assistance arrived, July 1st.

Sergt. Green, Company H, 13th Infantry, medal of honour. Conspicuous coolness and steadiness in handling his piece under hot fire, July 1st.

Sergt. John Graham, 10th Cavalry, medal of honour. Conspicuous coolness and steadiness under fire, July 1st.

Sergt. Weischaar, Company A, 13th Infantry, certificate of merit. Particularly meritorious steadiness, night of July 6th. Being put on outpost duty with a Gatling gun in time of truce, and having been alarmed by a sentinel, whose duty it was to warn him of the enemy's approach, he coolly reserved his fire for personal investigation and prevented a violation of the truce.

Sergt. Ryder, Company G, 13th Infantry, certificate of merit. Particularly meritorious steadiness, night of July 6th. Being on outpost duty with a Gatling gun in time of truce, and having been alarmed by a sentinel, whose duty it was to warn him of the enemy's approach, he coolly held his fire for personal investigation and prevented a violation of the truce.

In making these recommendations, I have limited myself to those which I personally observed. If I recommended for every deserving act, there is not a man in my whole detachment who has not deserved a certificate of merit. They were selected in the beginning from an army corps for what I knew of them, and they have abundantly justified my confidence in them. With a less efficient personnel it would have been absolutely impossible to organize, equip and instruct the first battery of Gatling guns ever used in the history of war, in the short space of time allotted me, and put it in efficient fighting shape. They fought their guns on the skirmish line and in advance of it, standing boldly up to do it when the skirmishers themselves lay down close for cover. My loss, as footed up on the night of July 1st, was 33 1-3 per cent, killed, wounded, and missing. The efficiency of the work of my guns was attested to me by numerous Spanish officers and prisoners. Their favourite expression was: "It was terrible when your guns opened, always. They went *b-r-r-r*, like a lawn mower cutting the grass over our trenches. We could not stick a finger up when you fired without getting it cut off—so!"

The work of this experimental battery proves that in this weapon we have a new arm supplementary to infantry and cavalry, independent of both as one arm is of another, and more nearly capable of independent action than any other arm of the service. It is equally demonstrated that this new arm is entirely different from artillery in its functions, and can live where the latter is compelled to retire.

It should, therefore, be organized as a separate arm. I have, at the request of General Wheeler, drawn up a scheme of such an organization and submitted it to him.

Experience shows me that the carriage is too heavy. I can only renew the representations contained in my letter of January 1, 1898, to the Adjutant General, accompanying drawing, etc., of my proposed carriage for machine guns. I would now, based on experience, modify my theory of organization as then proposed, and would make several changes in the model of carriage then proposed without departing from the general principles.

If any expression of such views is desired, I shall be very glad to submit them when called upon by the War Department to do so.

Very respectfully,

John H. Parker

2nd Lieut., 13th Infantry

Commanding Gatling Gun Detachment, 5th Corps.

LEONAUR

ALSO FROM LEONAUR
AVAILABLE IN SOFTCOVER OR HARDCOVER WITH DUST JACKET

CAPTAIN OF THE 95th (Rifles) *by Jonathan Leach*—An officer of Wellington's Sharpshooters during the Peninsular, South of France and Waterloo Campaigns of the Napoleonic Wars.

THE KHAKEE RESSALAH *by Robert Henry Wallace Dunlop*—Service & adventure with the Meerut volunteer horse during the Indian mutiny 1857-1858

BUGLER AND OFFICER OF THE RIFLES *by William Green & Harry Smith* With the 95th (Rifles) during the Peninsular & Waterloo Campaigns of the Napoleonic Wars

BAYONETS, BUGLES AND BONNETS *by James 'Thomas' Todd*—Experiences of hard soldiering with the 71st Foot - the Highland Light Infantry - through many battles of the Napoleonic wars including the Peninsular & Waterloo Campaigns

A NORFOLK SOLDIER IN THE FIRST SIKH WAR *by J W Baldwin*—Experiences of a private of H.M. 9th Regiment of Foot in the battles for the Punjab, India 1845-46

A CAVALRY OFFICER DURING THE SEPOY REVOLT *by A.R.D. Mackenzie*—Experiences with the 3rd Bengal Light Cavalry, the Guides and Sikh Irregular Cavalry from the outbreak to Delhi and Lucknow

THE ADVENTURES OF A LIGHT DRAGOON *by George Farmer & G.R. Gleig*—A cavalryman during the Peninsular & Waterloo Campaigns, in captivity & at the siege of Bhurtpore, India

THE COMPLEAT RIFLEMAN HARRIS *by Benjamin Harris as told to & transcribed by Captain Henry Curling*—The adventures of a soldier of the 95th (Rifles) during the Peninsular Campaign of the Napoleonic Wars

THE RED DRAGOON by *W.J. Adams*—With the 7th Dragoon Guards in the Cape of Good Hope against the Boers & the Kaffir tribes during the 'war of the axe' 1843-48

THE LIFE OF THE REAL BRIGADIER GERARD - Volume 1 - THE YOUNG HUSSAR 1782 - 1807 *by Jean-Baptiste De Marbot*—A French Cavalryman Of the Napoleonic Wars at Marengo, Austerlitz, Jena, Eylau & Friedland

THE LIFE OF THE REAL BRIGADIER GERARD Volume 2 IMPERIAL AIDE-DE-CAMP 1807 - 1811 *by Jean-Baptiste De Marbot*—A French Cavalryman of the Napoleonic Wars at Saragossa, Landshut, Eckmuhl, Ratisbon, Aspern-Essling, Wagram, Busaco & Torres Vedras

www.ingramcontent.com/pod-product-compliance
Lightning Source LLC
Chambersburg PA
CBHW032056080426
42733CB00006B/304